INTERIOR PLANNING AND DESIGN

Interior Planning and Design: Project Programs • Plans • Charettes
Christina M. Scalise

Vice President, Technology and Trades SBU:
Alar Elken

Editorial Director:
Sandy Clark

Senior Acquisitions Editor:
James Devoe

Senior Development Editor:
John Fisher

Marketing Director:
Cyndi Eichelman

Channel Manager:
Fair Huntoon

Marketing Coordinator:
Sarena Douglass

Production Director:
Mary Ellen Black

Production Manager:
Andrew Crouth

Production Editor:
Stacy Masucci

Art and Design Specialist:
Mary Beth Vought

Technology Project Specialist:
Kevin Smith

Editorial Assistant:
Mary Ellen Martino

Cover images courtesy of: Shelley Jurs, Jurs Architectural Glass and Milroy and McAleer Photography (Valley Hilton Lobby).

Library of Congress Cataloging-in-Publication Data:
Card Number: [Number]

ISBN: 1-4018-2809-4

NOTICE TO THE READER

INTERIOR PLANNING AND DESIGN

project programs • plans • charettes

Christina M. Scalise

Contributing authors:

Maximilian J. Mavrovic

Maureen Vissat

THOMSON
™
DELMAR LEARNING

Australia   Canada   Mexico   Singapore   Spain   United Kingdom   United States

DEDICATION

Dedicated to AR and DV.
In memory of Catherine and Clara.

CONTENTS

Section 3   Retail Facilities

Section 4   Hospitality Facilities

Section 7   Historical Facilities

Section 8   Medical Facilities

Section 9   Personal Service Facilities

Section 10 Exhibit Facilities

Section 11 Submittal Resources

PREFACE

In order for natural creativity to flow spontaneously, the design process involved must become second nature. The communication of the creative process is fairly systematic, and when clarified and practiced, it is a highly valued skill for bringing creativity into the light. *Interior Planning and Design: Project Programs, Plans, and Charettes* is a core teaching resource with projects and exercises intended to assist in providing students the experience and practice to effectively develop and communicate interior design concepts. The varied design projects are comprehensively presented, with authentic client programs. The organization of each project and the subsequent submittal requirements set into practice the phases of the design process. This feature allows the book to be used throughout the entire design program and at every student level. You may wish to assign only the preliminary design segment of the project if you are teaching programming and planning to the students. One of the great benefits of the chapter arrangement is that these revisions are simple to implement and is the intention of the initial text concept.

The book is organized according to Facility Types, including Educational, Office, Retail, Hospitality, Recreation, Residential, Historical, Medical, Personal Service, and Exhibit. The projects vary in square footage and submittal requirements to provide a range appropriate for second-year through fourth-year-level students. The vast selections of projects are intended to be used at different course levels and include but are not limited to a few feature exercises for metric application, special uses, projects integrating a team approach, and several charrettes of short time limits. Included are various project forms, schedules, and evaluation documents. The Project Reference Table illustrates the project facility types, sizes, and design phase coverage included in this book, listed by chapter on the top row of the table.

There are an unlimited amount of project options and variations that can be applied to each chapter. If instructors consider a project difficult for the level of a particular student group, the submittal requirements could be revised, or perhaps the assignment could be altered to include one or two spaces of the project.

Opportunities for one-room projects could be selected from numerous chapters. The Logo Shop in Chapter 1, the Charles Jourdan Department in Chapter 14, and the Lobby or Restaurants in Chapter 16 could each be a project in itself. And of course, the ever-challenging kitchen is expanded to be a baking and catering kitchen in Chapter 25. The Grille Room or Ballroom in Chapter 15 could be singled out for client interviews and research. Chapter 23 could offer a one-room home theater project, and Chapter 25 a Safe Room and Shelter Room.

Reviewer Thomas Houser has suggested the following possibilities in his evaluation of the book. While assigning the residential project for a client with a physical challenge, assign a different handicap for each student. Likewise, assign different menus or average tabs on the restaurant projects. The Tango Club is an excellent project that could be tailored for studios at various levels within the curriculum because there is a lot of information to cover, from image to store fixtures. The conference table design requirements in Chapter 3 could be expanded into a good project for a furniture design class. The development of the graphics in Chapters 8, 16, 17, and 30 could offer opportunities for team projects involving interior design and graphic design students. Chapter 2 could be a good project for a business procedures class, as well as one for a studio.

PROJECT REFERENCE TABLE

Project Chapter	1	2	3	4	5	6	7	8	9	10	11	12	13	14	15	16	17	18	19	20	21	22	23	24	25	26	27	28	29	30
Design Phase Outcome																														
Pre Agreement	•																													
Programming	•	•	•	•	•	•	•	•	•	•	•	•	•			•	•	•		•	•	•	•	•	•			•	•	•
Conceptual Design		•	•	•	•		•	•	•		•	•	•			•	•			•		•	•	•	•			•	•	•
Design Development			•	•					•	•	•	•	•	•						•	•	•	•	•	•			•	•	•
Construction Documents		•	•						•											•	•	•								•
Model Presentations																														•
Code Meeting							•							•																
Charette					•					•			•						•											
Team Project							•							•																
Facility Type																														
Educational	•																													
Office		•	•	•	•	•	•																							
Retail								•	•	•	•	•	•	•																
Hospitality															•	•	•													
Recreation																		•												
Residential																			•	•	•	•	•	•	•					
Historical																										•	•			
Medical																												•		
Personal Service																													•	
Exhibit																														•
Special Issue Addressed					•		•							•						•		•		•	•					
Square Footage																														
1000 or less	•	•													•				•											
5000 or less					•	•		•	•	•	•	•	•			•	•			•	•	•	•			•	•	•	•	
5000 and above			•	•			•							•				•						•	•					•

Special issue includes ecological, security and safety or accessibility.

Chapters 1 through 30 consist of individual student projects and are organized in the following format:

Outcome Summary: The outcome summary describes the design phases included for the project and the unique focus of the exercise.

Client Profile: This section describes the client and includes an overview of the particular client's requests.

Project Details: A detailed description of the project program developed from client interviews is included. The subheadings are:

- General and Architectural Information, which lists location, square footage, project budget, and so on;
- Codes and Governing Regulations, which specifies codes to use. However, the instructor may reassign with local codes or the International Building Code;
- Facilities Required; and
- Spatial Requirements, which are organized per room, department, or area with planning, design, and furniture and equipment requirements. Details are specific (sizes and quantities) where the client has requested. There are also projects that require thorough student research and use of previously learned methodology.
- Floor plans and related drawings and photos are provided as additional references for the projects.

Submittal Requirements: The submittal requirements take the student through the stages of the design process, and include a due date for each phase so the student can learn to budget his or her time more efficiently. For the same purpose, as well as supporting the student in project management skills, included with each project is a requirement to prepare a student time management schedule.

Each submittal phase includes a suggested evaluation document that can be found in Chapter 31. They are structured relative to the scope of services of interior design, and are expanded to include detailed design process tasks for each phase. The general evaluation format is a point system for each task. It is based on typical jury sheets used for student design competitions. The point allocation is to be determined by the design educator.

AutoCAD LT 2002 Drawing File Name: This is a drawing (.dwg) file of the project that is included on the student CD supplement. Many of the projects also include a ceiling plan and elevations. The plans and elevations assume that the students have a basic knowledge of computer-aided drafting.

References: Textbook support for design process teaching and books related to project type for further reading are listed at the end of most chapters. Typical resources for a facility type are listed in the first chapter of each section. The recommended books, organizations and, trade magazines contain information that will be useful in completing the projects

Special Resources: Students are quite fortunate to begin their learning at a time when an infinite amount of resources are literally at our fingertips. Web-related trade resources specific to the project are listed here to start a research journey.

Product Manufacturers: Web-related trade resources for furniture, fabrics, finishes, fixtures, and equipment are included. The product manufacturers (interior trade resources) listed at the end of each project are starting points for research, and establish project quality and price points. So there is no excuse for cutting a picture of a magazine rack or task light from a mail-order or retail sales catalog, as many students are tempted to do.

Notes: An area is provided for student notes or further instructions from the educator.

Chapter 31, Submittal Documentation, contains the evaluation documents for Programming, Conceptual Design, Design Development, Final Design Presentation, Construction Documentation, Design Agreements and Proposals, and Exhibit Design and Model Presentation.

Support information and tools are also included in Chapter 31. The CAD Layer Designations are discussed and defined with two tables. The forms required for completing assignment elements are provided here and referenced to the student CD where applicable. The forms include a Project Record Sheet, Furniture Specification, Color and Material Documentation, Purchase Order, Invoice, Sample Matrix, and a Time Management + Project Schedule Form. It should be a requirement for the students to prepare this schedule for each project, excluding the project charettes. A few typical retail drawings are included. Governmental regulations, the ADA, and building code discussion and resources are also listed in this chapter. Remember: licensed professionals, such as architects, must determine code requirements.

A Supplement Instructor's CD includes Microsoft PowerPoint Presentation files for each chapter. Use them for the instructor's initial project reviews with the student. The chapter presentation files include highlights from each chapter, project photos and drawings. Solutions are also part of this tool and may include preliminary sketches, plans, renderings, color boards, construction documents, and final project photographs. I recommend showing the solutions to the students after they make their final presentations.

about the author

Christina M. Scalise, principal of her own firm, is a professional interior designer whose interior and product design work has been published often in trade magazines and international publications. In practice since 1975, her hospitality and retail design work has won numerous awards and honors. While employed with Cole Martinez Curtis & Associates, she served as vice president, design director of the retail division and as a member of the Executive Management Group. She was also the design director of the Interior Design division of MCG Architects, before starting her own practice in Newport Beach, California, Florida, and Pennsylvania. As a part-time interior design educator, Scalise taught design studio, professional practice, and rendering courses at Woodbury University, Seton Hill University, Brooks College, and the Art Institutes beginning in 1982. She has also served on the board of the Institute of Store Planners and as the director of its National Student Competition. Scalise is a graduate of Syracuse University where she received a Bachelor of Fine Arts degree. She is certified by the National Council for Interior Design Qualification and is affiliated with several professional organizations.

contributing authors

Maximilian J. Mavrovic, AIA, is the principal of Mavrovic Architects PC. Before establishing his architectural design firm, Maximilian J. Mavrovic served as vice president at two of the largest architectural firms in Pittsburgh, where he was directly responsible for client relations, staff supervision, and project design and budgets. Mavrovic is a graduate of Carnegie-Mellon University where he received a Bachelor of Architecture in 1983. His professional affiliations include the Pennsylvania Society of Architects and the National Council of Architectural Registration Boards

Maureen Vissat is an associate professor at Seton Hill University, Division of Visual and Performing Arts where she teaches art history and leads independent and group studies outside the United States to museums, art collections, and sites throughout Italy, Spain, Greece, France, England, and Canada. Her honors include CASE Professor of the Year, chair of the University's Core Curriculum Committee, and board member for the Allegheny Historic Preservation Society. Vissat received her M. A. in Art History from the University of Pittsburgh and a B.A. from Georgetown University.

acknowledgments

The Delmar Learning team, including James Devoe, Senior Acquisitions Editor, John Fisher, Senior Development Editor, Sarena Douglas, Marketing Coordinator, Mary Beth Vought, Senior Art/Design Coordinator, Stacy Masucci, Production Editor, Mary Ellen Martino, Editorial Assistant. Thank you for the privilege and the composed and expert coaching.

I appreciate the reviewers rich critiques that are now laced inside the text. LuAnn Nissen, University of Nevada, Reno, NV, Thomas L. Houser, Lamar Dodd School of Art, Athens, GA., and Wen Andrews, J. Sargent Reynolds Community College, Richmond, VA..

For the pure kindness and support of my family, especially Anthony, Dolores, Michael, Mark, Tony, Joe, Debbie, Rob and Anne. To such gifted associates as Max, Maureen, Guy, Christopher White, Reyerson Tull Corporation, Hal, Steve and Gary Stearns, Tom and Susan Certo, Ibacos Incorporated, Stacy Hunt, Glen Cotrell, Scalise Industries, Brud Bavera, PSR Associates, Katherine M. Barley, The Pittsburgh Music Academy, Dr. Dale DeConcilis, Ed and Gail Wayne, MCG, Rick Gaylord, May D&F, Cole Martinez Curtis and Associates, Dennis Takeda, Carl Turnbull, Durant/Flickinger Associates, Robert Pisano, The Agree Corporation, "anonymous client," Jack Lee, Miss Yeh, Mr. Le, Sherman Chiang, MCA, Vicky and Steve Bates, Shelley Jurs, Wolf Major, MGA, Kevin Klein, Ernie Pope, Louis Aguilar, Ken Smith, Elaine Restauri, Frank Rapp, Don Conway, Venetia and Gary Torre, Seton Hill University, April Maskiewicz, Charles Wilson, John Sotirakis, Leo Martinez, Al Cruz, Bob McClellan, Allan Hing, Pat Kooser-Wall, Francis Krahe, Jin Au, Eva, Carol Tivoli, Christianne Hillery, Heather Clark, Barbara Moore, Susan Hahner, Anne Hoffmann, Danielle Raishart, Jessica M. Lynch, Charlie, Don and Vern, The Rivers Club, Holly, Joe, Bill Pavelitz, Diane Minn, Debbie Halferty, Joe Magnetti, Carol and Jeff, JoAnne Kopp, Colleen Miller, William Wong, KMA & Associates, Jayne Peterson, Jim Snyder, Uncle Phil, Susan, Mary Knox-Henderson, Ramona Escano, Cindy Rubin, Kurt Cyr, Blake DeMaria, Matt, Alex, Lynne, Masako, Tammy, Kristen, Bo, Dewayne Neufeld, Rick, Nino Nieman, Alice, Micki, Reiko, Carolyn, Becky, Debbie, Ernie, Gary, Cecelia, Tiodi, Brook Rege, Lily Tan, Cheryl, Bruce Flooring, Armstrong Corporation, Alistair Mackintosh Ltd., Michael Nunokawa, Murayama Kotake Nunokawa & Associates Inc., Hawii Omori Corporation, Harold Mizomi, Mel Kitagawa, Teknion LLC, and W. Wyatt Neel.

I am truly grateful.

Christina M. Scalise

You may contact the author through Delmar Learning at http://www.delmarlearning.com.

EDUCATIONAL
FACILITIES

OPEN
BALCONY

THIRD
FLOOR

OPEN
BALCONY

SECOND
FLOOR

FIRST
FLOOR

CHAPTER 1

The Pittsburgh Music Academy

OVTCOME SVMMARY

This project is an exercise that covers the design process, from planning and design, through design presentation. The size of the space is large and the presentation requirements are extensive. The preliminary submission requirements are in stages, and the final presentation requirements are comprehensive, including renderings and budgets.

client profile

The Pittsburgh Music Academy provides aspiring young artists with musical instruction necessary for instrumental performance. The academy specializes in teaching young children, starting at age three, by using the Suzuki method. Mr. Charles Wilson tells us that the school believes the core of music instruction is the time spent by the teacher and the student on a one-to-one basis. During these lessons the student is taught the structure and form of music in addition to learning to play music with others. The school's Suzuki philosophy creates a relationship triangle composed of the teacher, the student, and the parent. Mr. Wilson explains that as the student matures and progresses through the various levels of instruction, the parent's role changes from an active coach to a supportive one.

The PMA has retained you to plan and design the new facility. Wilson is facilitating the design process. Ms. Katherine Barley is the academy director and CEO. Barley handles all of the academic issues and supervises the faculty. She is extremely well-known for her accomplishments as a musician and teacher of the Suzuki method. The secretary of the corporation is Ms. Carolyn Hills. There are currently eleven faculty members, including Jeffrey Turner, who is the PMA's orchestra conductor, and Jennifer Moe, chair of the Violin Department. Barley is proud to say that the faculty is composed of a very fine group of well-trained, energetic, and caring musicians.

Full matriculation in the PMA program requires participation in private lessons, music education, ensembles, recitals, and concerts. Partial matriculation requires participation in only one of those features. This year the schedule offers classes in Private Lessons, Cello Group, Violin Group, Viola Group, Flute Group,

Figure 1-1
PMA Logo Design

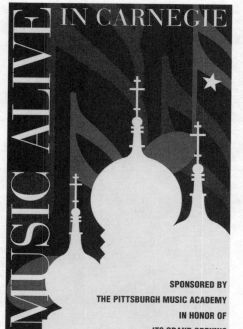

Figure 1-2   Opening Concert Poster by John Sotirakis. Concept to be used for sign banners.

Music Games for Piano and Flute, Orchestra, Piano Groups, Theory I and II for Piano and Flute, Show Choir, Chime Choir, Music Together, Adult Piano Group, Advanced Piano Group, Pre-Twinkle String Group, Theory I and II for Strings, Music Games Class for Strings, Intermediate and Advanced Chamber Ensembles, Music History, and Advanced Theory. Currently, there are 150 students enrolled in the school. The maximum number of students who would occupy the building at one time is sixty, and that only occurs on Saturdays. In five years PMA expects the enrollment to grow to 390 students. At that time, an additional site would open.

The PMA will be leasing a building at 232 Third Avenue in Carnegie, Pennsylvania. It is on the corner of Third Street and Third Avenue, a good omen because the famous Settlement Music Schools in New York City and Philadelphia are both on Third Street! The school chose Carnegie as its home base because of its accessibility from all directions and its reputation as one of the "100 Best" small arts towns in the country. The academy looks forward to working along with other Carnegie businesses and corporations to support the local arts council and the music program at the Boys and Girls Club. The landlord of the building is a local business owner in the construction industry and has graciously agreed to build out the space according to PMA's needs. The firm of Mavrovic Architects has been hired to complete the architectural portion of the project.

Mr. Wilson has secured the services of John Sotirakis of AMS for the graphics and signage. John has been working with PMA since its inception and has developed the PMA logo, letterhead, exterior signage, and newsletter. The color palette and how the interior signs are to be integrated are parts of your scope for this project.

project details

General and Architectural Information

Location: The Pittsburgh Music Academy, 232 Third Avenue, Carnegie, Pennsylvania

Gross square footage: 5,568 gross building area. Four levels: 1,392 square feet

Bidding required: Landlord to act as general contractor

Opening or move-in date: October

Consultants required: Architect, electrical engineer, and mechanical engineer

Codes and Governing Regulations

At the time of construction, the Pennsylvania Labor and Industry Department classified this building as Ordinary Type Construction with a D-O Occupancy Group. The Boca 1996 Building Codes and the PA L & I both require a minimum of two exits from the building and exit signs throughout. The codes require a fire extinguisher on each level. The landlord will be installing automatic sprinklers, although code does not require them. A disability restroom facility is required, and code also requires four water closets and four lavatories for the building. A janitor closet with a sink is required on each level. You must abide by the most current basic requirements of the above-noted standard regional building codes, fire and safety codes, and the Americans with Disabilities Act.

Figure 1–3
Exterior of building before the renovation

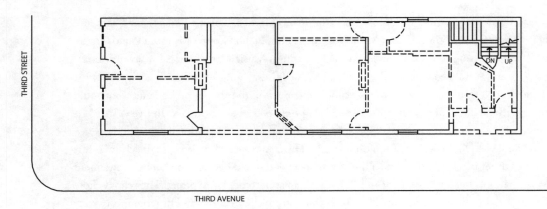

Figure 1–4   First-Level Floor Plan. Ceiling height is 9'3". Dashed lines indicate walls to be demolished.

First Level Spatial Requirements

ENTRY

Planning

- The entry will be the first space encountered by those entering and should reflect the spirit of PMA. A sign that identifies PMA and a bulletin board to post class schedules and memos will be needed.
- PMA will have a volunteer "greeter" on duty during the afternoon, evenings, and all day Saturday. PMA prefers that this person have a place to sit right inside the entry doors to watch as parents are dropping off the children for their lesson. This greeter also requires access to the intercom system to help students and parents find each other in the building if necessary.

Furniture

- A podium-type desk with locked storage for rain gear, safety vest, and flashlight; a shelf for schedule books and student rosters
- Bar-height stool with a back and casters
- Bulletin board: 30" high × 36" wide

Equipment

- Telephone with an intercom
- Security devices will be installed in the facility, including access control devices and intrusion detection devices, which are tied into the local police authorities.

Design

- PMA requires a hard surface floor and movable floor mats.

WAITING ROOM

Planning

- The waiting room should be next to the entry and the receptionist. The room is used for visitors and for adults and children who are waiting for lessons to begin. During the weekend, complimentary coffee and pastries are served in this area. Provide additional small waiting areas on the second and third levels to accommodate the larger numbers of visitors and students on Saturdays.
- PMA requires storage of a modular type for coats, hats, umbrellas, and instrument cases (the instruments remain with the students). This storage may also be located in Entry as space allows.
- Provide a small play area for younger children in the waiting area.

Figure 1–5 CAD Elevation of the Architectural Revisions, Third Street

Figure 1–6 CAD Elevation of the Architectural Revisions, Third Avenue

Furniture
- Seating for at least twenty adults
- Side tables with lamps
- A magazine and book rack
- Serving table for coffee and pastries
- Seating for at least four children with play table and game and toy storage
- Artwork

Design
- PMA requires a hard surface floor near the entry door and movable floor mats.
- A commercial grade carpet for the waiting area is required.

RECEPTION/REGISTRAR AREA

Planning
- The reception area should be adjacent to the waiting area and near the entry. This area will be the workspace for the registrar during normal business hours. A volunteer receptionist will be on duty during the evening and weekend hours.
- The registrar will be enrolling the students and collecting tuition, processing correspondence, and performing additional tasks for three school administrators. The duties of the registrar also include answering the telephone and directing students or parents to their destination.
- The registrar will also be managing paperwork and performing the task of bulk purchasing for the logo shop and snack bar.

Furniture
- Desk-height work surfaces (approximately thirty six square feet), including a retractable keyboard tray and pencil drawer; a transaction counter work surface
- Two 36″ wide lateral file drawers with locks
- One mobile pedestal with a file and box drawer with locks
- Overhead storage of at least 72 linear inches
- A task chair with casters for carpet
- Small bulletin board 24″ wide × 30″ high
- Task light at the desk
- Artwork

Equipment
- Typewriter
- A computer with a printer
- Telephone with an intercom

Design

- The furniture can be a panel system.
- Commercial grade carpet is required.

KITCHEN

Planning

- The kitchen will be used to provide minimal food storage and preparation space for employees who will be in the building for long periods. Ten to twelve linear feet of counter and storage is sufficient.
- Caterers will prepare food here for small meetings and intimate, academy-related gatherings.
- The coffee and tea urns will be kept here.
- Three linear feet of storage for snack shop items is required.

Equipment

- Cold storage for food and ice
- Dishwasher
- Ice maker
- Sink
- Plumbing for a coffee maker
- Microwave
- Electrical receptacles at counter level

Design

- Hard surface flooring is required in this area.

SNACK BAR

Planning

- The snack bar will be a portable setup of approximately three linear feet that will allow the sale of prepackaged snack items to students and parents on Saturdays only. It should also allow for the sale of cold and hot drinks.
- The fixture can be stored in the kitchen, conference room, or basement if it is knockdown. The snacks will be stored in the kitchen during the weekdays. No special equipment is required.

LOGO SHOP

Planning

- The logo shop will sell PMA logo merchandise such as T-shirts, sweatsuits, coffee mugs, notepads, pencils, buttons, pens, posters, jewelry, transfers, and other novelty items related to the academy.
- PMA requires 4 linear feet of locked display for the small gift items. Ten linear feet is required for displaying the folded clothing merchandise and larger gift items.
- The success of the shop will depend on how the merchandise is presented and displayed and its location. It needs to be next to the waiting area for exposure and visible to the receptionist area for servicing and security reasons. A full cash wrap station is not required.
- Storage for back stock of merchandise is provided in the basement.

Equipment

- Fixtures as noted above with display lighting and power within the fixtures.
- Power for future POS (Point of Sale) terminal
- Storage for tissue paper and bags

ADMINISTRATION OFFICE

Planning

- The administration office space will be the "information brain" of the academy and provide workspace for the administrators and a conference space. Three separate workstations are required for Charles Wilson, Kathie Nee, and Kiki Barley.
- In this area you will need to create a virtual office space by providing two stations to be used by all faculty members. This area should provide for the distribution of mail and information and provide personal storage for each faculty member.
- The main work done in this space includes billing, correspondence, administrative meetings, individual conferences, scheduling, and academy-related business activities.
- File storage is required for all archives including student records and financial records.

Furniture

For administrators:

- Desk-height work surfaces (twenty five to thirty square feet), including a retractable keyboard tray and pencil drawer
- Two 36″ wide lateral file drawers with locks
- One mobile pedestal with one file and one box drawer with locks
- At least 60 linear inches of overhead storage, 12″ deep
- A task chair with casters for carpet
- A guest chair
- A tacking surface
- Task light at the desk
- Artwork

For the faculty:

- Two work surfaces of approximately five square feet each
- Eleven small storage cabinets with locks that will serve as a faculty lockers, with a mobile pedestal with a file and box drawer with locks for each storage locker
- Mail distribution system for eleven persons
- Two task chairs with casters for work surface spaces
- A tack surface
- Artwork

For general area:

- Storage shelving with doors, a minimum of 60 linear inches, 12″ deep
- Locking file storage, a minimum of 240 linear inches
- A conference table to seat six to eight people
- Six to eight conference chairs sized appropriately for a limited amount of space
- Presentation board with marker surface, projection screen, and tackable panels
- Artwork

Equipment

- Fax machine
- A networked printer
- Computer terminal at each administrator's workstation
- Telephone with an intercom at each workstation

Design

- Commercial grade carpet is required.
- Furniture should be modular and provide privacy at each station. Both free-standing and panel systems or any combination of may be used
- The door into the office should have partial glass.

STUDIO A

Planning

- This studio will be used for teaching on weekdays, and should be designed for Suzuki private lessons that accommodate the student, the teacher, and the parent.
- The space will also be used as an additional activity space for students waiting for lessons.
- A space of 80 to 100 square feet will be sufficient.

Furniture

- Two music stands with lights
- Storage and bookcase for two linear feet of music materials and textbooks
- Bulletin board and marker board: 36″ high × 48″ wide
- Three chairs
- Desk or table for the instructor

Equipment

- Telephone with an intercom
- A portable electric keyboard
- Upright piano with bench

Design

- Commercial grade carpet is required.
- The door into the studio should have partial glass.

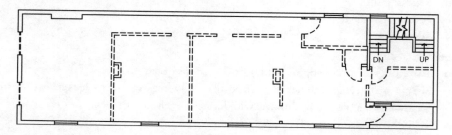

Figure 1–7   Second-Level Floor Plan. Ceiling height is 8′10″.
Dashed lines indicate walls to be demolished.

Second Level Spatial Requirements
STUDIO B

Planning

- This studio will be used for teaching on weekdays and Saturdays, primarily for flute and string lessons.
- The space will also be used as an additional activity space for students waiting for lessons when the orchestra is practicing in the rehearsal room.
- A space of 80 to 100 square feet will be sufficient.

Furniture

- Two music stands with lights
- Storage and bookcase for two linear feet of music materials and textbooks
- Bulletin board and marker board: 36″ high × 48″ wide
- Three chairs
- Desk or table for the instructor

Equipment

- Telephone with an intercom

Design

- Commercial grade carpet is required.
- The door into the studio should have partial glass.

REHEARSAL ROOM 1

Planning

- This will be the main rehearsal space, and it will accommodate large rehearsals for the orchestras, chamber music, choir, recording sessions, music history, and advanced theory and composition classes.
- When not in use as a teaching space, the room will be scheduled for monthly graduations, faculty meetings, and parent meetings.
- This space will also function as the listening lab, computer lab, and the music library. A separate room is not required for these facilities. Appropriate fixtures using wall space will be adequate.
- Private lessons will be taught here, and it will also be used as a waiting space for students and parents when not in use.

Furniture

- Thirty-five to forty music stands with lights
- Thirty-five to forty stacking, folding tablet-arm chairs
- Bulletin board and marker board: 48″ high × 72″ wide, with a projection screen
- A conductor's podium with stool and chair
- One desk or table for the instructor and librarian
- Seating for three to five observers
- Music library modular storage system, including ability to store sheet music and recorded music
- Three listening lab/computer synthesizer workstations, 4′ × 2′-6″, including chairs

Equipment

- Telephone with an intercom
- Three computers
- Stereo equipment with ability to play records, tapes, and CDs
- An overhead projector
- Television and VCR on a mobile stand
- Internal recording equipment
- 7′ grand piano
- Mobile storage carts for student chairs and music stands
- Portable riser platforms for the orchestra and choir

Design

- The flooring should be a hard surface material for acoustics and easy arrangement of equipment.
- Acoustical panels are required and designed in such a way that they can be applied to the wall surfaces at the corners of the room or a portable system such as acoustical shells as manufactured by Wenger Corporation. Drywall finished walls and ceilings are required because acoustical tile ceilings will absorb too much of the sound.
- Indirect general illumination is preferred in combination with direct incandescent lighting, which is better for reading music.
- The door should have partial glass.

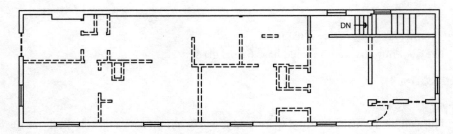

Figure 1–8
Third-Level Floor Plan. Ceiling height is 8'6". Dashed lines indicate walls to be demolished.

Third Level Spatial Requirements

STUDIO C

Planning
- This studio will be used for teaching on weekdays and Saturdays, and should be designed for Suzuki private lessons that accommodate the student, the teacher, and the parent.
- A space of 80 to 100 square feet will be sufficient.

Furniture
- Two music stands with lights
- Storage and bookcase for music materials and textbooks
- Bulletin board and marker board: 36" high × 48" wide
- Three chairs
- Desk or table for the instructor

Equipment
- Telephone with an intercom

Design
- Commercial grade carpet is required.
- The door into the studio should have partial glass.

STUDIO D

Planning
- This studio will be used for teaching all private instrument lessons and small group classes.
- A space of 80 to 100 square feet will be sufficient.

Furniture
- Two music stands with lights
- Storage and bookcase for music materials and textbooks
- Bulletin board and marker board: 36" high × 48" wide
- Three chairs
- Desk or table for the instructor

Equipment
- Telephone with an intercom
- Upright piano

Design
- Commercial grade carpet is required.
- The door into the studio should have partial glass.

REHEARSAL ROOM 2

Planning

- This will be the main rehearsal space for small chamber rehearsals, music theory classes, and for private lessons when not in use as a classroom.

Furniture

- Fifteen music stands with lights
- Fifteen stacking, folding tablet-arm chairs
- Bulletin board and marker board: 48″ high × 72″ wide with a projection screen
- Conductors' podium with stole and chair
- Desk or table for the instructor
- Seating for five observers
- Storage and bookcase for music materials and textbooks

Equipment

- Telephone with an intercom
- Stereo equipment with ability to play records, tapes, and CDs
- 7′ grand piano
- Mobile storage carts for student chairs and music stands

Design

- The flooring should be a hard surface material for acoustics and easy arrangement of equipment.
- Acoustical panels are required and designed so that they can be applied to the wall surfaces at the corners of the room or a portable system such as acoustical shells as manufactured by Wenger Corporation. Drywall finished walls and ceilings are required because acoustical tile ceilings will absorb too much of the sound.
- Indirect general illumination is preferred in combination with direct incandescent lighting, which is better for reading music.
- The door should have partial glass.

SUBMITTAL REQUIREMENTS

Phase I Programming   Date Due: _____

A. Research various shapes, sizes, arrangements, and clearances for the above outlined spaces. Research furniture and equipment for the same.

B. Analyze the program outlined above and develop a programming matrix for each level keyed for adjacency locations. A sample blank matrix is provided in Chapter 31.

C. With the information gathered from the matrix, provide a minimum of three bubble diagrams for each level, showing the spatial relationships of the areas required. Provide a one-page typed concept statement.

Presentation formats for Parts A through C: White bond paper, 8½″ × 11″, with a border. Insert in the binder or folder. The matrix and bubble diagrams are to be completed on the CADD program.

Evaluation Document: Final Design Presentation Evaluation Form

Phase II: Planning and Preliminary Design

Part A Due: _____   Part B Due: _____   Part C Due: _____

A. Provide preliminary partition plans showing all walls, ceilings, built-in fixtures, bathroom fixtures, kitchen fixture layout, doors, and windows. Label rooms. This plan can be a freehand drawing or CADD drawing in 1/4″ scale. Drawings should have an appropriate border. Finished drawings should be blueprint or white bond copy.

B. Develop the above partition plans to include a preliminary ceiling, furniture, and furnishings layout of each room. Label rooms. This plan can be a freehand drawing or CADD drawing in 1/4″ scale. Drawings should have an appropriate border. Finished drawings should be blueprint or white bond copy.

C. Develop the above furniture plans into final floor plans showing all items noted in the program. This plan is to be completed using the CADD program. Develop ceiling and lighting studies into final reflected ceiling plans, showing all ceiling conditions and heights on the CADD program.

Evaluation Document: Conceptual Design Evaluation Form

Phase III: Preliminary and Final Design

Part A Due: _____   Part B Due: _____   Part C Due: _____

A. Develop a color, material, and final furniture scheme. Samples and pictures can be presented in a loose format.

B. Develop a perspective sketch of the waiting and reception area, showing all design elements and furnishings.

C. Make lighting fixture selections and key them to a legend, the reflected ceiling plan, and photos of the lighting fixtures. Photos can be presented in a loose format.

Evaluation Document: Design Development Evaluation Form

Phase IV: Final Design Presentation   Date Due: _____

A. The final presentation will be an oral presentation to a panel jury. Project must be submitted on 20″ × 30″ illustration boards. Your name, the date, the project name and number, the instructor's name, and the course title must be printed on the back of each board in the lower left-hand corner. Submit as many boards as required, including furniture plans, reflected ceiling plans, a colored elevation of one wall of a typical studio and one wall of a typical rehearsal room, a rendered perspective sketch of the waiting and reception area, signage concept study elevations, furniture selections, and material selections.

B. The plans should be presented in 1/4″ scale. The elevations are to be presented in 1/2″ scale. Illustration board size can be adjusted to fit the plan size. All boards must be the same size.

Evaluation Document: Final Design Presentation Evaluation Form

Phase V: Budget   Date Due: _____

A. Prepare the final budget, listing the cost of each furniture item per room. Present in typed format on 8½″ × 11″ white bond paper. Insert in the binder or folder.

Evaluation Document: Construction Documentation Evaluation Form
Project Time Management Schedule: Schedule of Activities
The AutoCAD LT 2002 drawing file name is pma.dwg and can be found on the CD.

REFERENCES

Book

Ballast, David Kent. 1994. *Interior construction and detailing for designers and architects.* Belmont, Calif.: Professional.

Special Resources

Amadeus Performance Equipment, Ltd.: http://www.amadeus-equipment.co.uk
Wenger Corporation: http://www.wengercorp.com

Product Manufacturers

DesignTex: http://www.dtex.com
Grand Rapids Chair Company: http://www.grandrapidschair.com
Kartell: http://www.kartell.com
Lees Carpets: http://www.lees-carpets.com
Loewenstein: http://www.loewensteininc.com
Office Specialty: http://www.officespecialty.com
The Amtico Company: http://www.amtico.com
Vecta: http://www.vecta.com

NOTES

NOTES

OFFICE FACILITIES

BATHROOM

CLOSET

OFFICE

CHAPTER 2

Executive Office

OUTCOME SUMMARY

This project is an executive office space, and its purpose is to gain proficiency in completing every phase of a project. It is divided into two parts. The first part of the project covers the planning and design phase. The second part encompasses construction documentation, including all plans, as well as furniture specifications, purchase orders, invoices, and a post occupancy questionnaire. The complete submittal requirements are extensive; the size of the space is a manageable 300 square feet.

client profile

William A. Wood and Sons, Inc. is a national distributor of metal products, such as steel and aluminum, and has been in business for over 50 years. Mr. Mark Dalton, the general manager of the distribution center located in Philadelphia, Pennsylvania, has hired you to design his office space. The duties of the general manager include overseeing the finances, sales, manufacturing, and operations of the company. Dalton uses his office to meet with clients as well as his staff.

Dalton is very proud of the many sales, quality, and safety awards that this distribution center has received, and loves to tell the stories of the long history of the company. He has on display many treasures of old company catalogs and metal models of the company trucks from the earlier days. A native of Chicago, Dalton owes his love of architecture and knowledge of design to his father. He developed a sophisticated eye for design while he was growing up, because his father was in the design profession. Dalton is in his mid-thirties and married with two children, whom he loves to share hockey games and camping trips with. Dalton is very warm and friendly and has a great sense of humor.

The building was originally built in 1945. The walls and ceilings are plaster finish over furring strips, and the floor is concrete slab construction. You may core drill to locate new power or communications outlets.

project details

General and Architectural Information

Location: William A. Wood and Sons, Inc., Philadelphia, Pennsylvania. Executive office renovation for Mr. Mark Dalton

Gross square footage: 308

Project budget: Funding for the project has been committed from the corporate capital expenses this year, and the budget allows for an expense of $40,000.00. The furniture, floor covering, lighting, electrical, and new drywall work will be sent out to bid by the owner; therefore, you will be required to submit furniture specifications as part of the construction documents.

Existing construction: The current facility will be renovated. Current facility plans are available. Site information gathering is required.

Consultants required: Architect, electrical engineer, and mechanical engineer

Codes and Governing Regulations

Occupancy: B – Business

Fire suppression: Standpipe and Hose System

Design the project to comply with the most current edition of the BOCA National Building Code (NBC), the National Electrical Code (NEC), the Uniform Fire Code (UFC), and the Americans with Disabilities Act for office spaces.

The owner and contractors will apply for any permits required for the construction work.

Facilities Required

OFFICE

Planning

- The office is used for the daily tasks of management, client meetings, and weekly meetings with six upper management staff members. Larger staff meetings are held in the conference room. Dalton spends the majority of his time on the telephone, meeting with clients, and reviewing data on his computer. His door is always open and staff members frequently meet with him on a daily basis in addition to the regularly scheduled staff meetings. He has requested easy access to the computer printout files that he refers to daily and would like them to be stored in his desk.

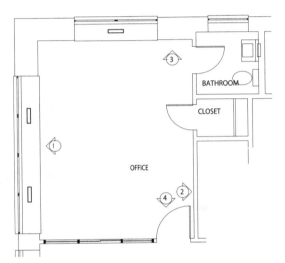

Figure 2–1 Floor Plan

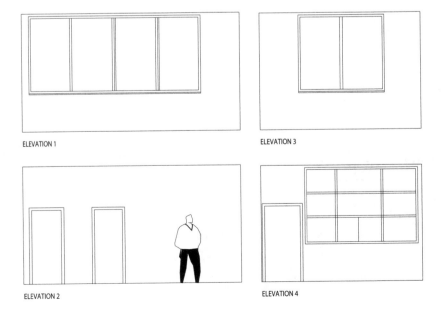

Figure 2–2 Elevations

- Dalton prefers an informal gathering with sofas and chairs for clients and staff meetings. When meeting with clients, blueprints are often reviewed, so an appropriately sized table of at least 36″ in length is required. There are many awards, both plaque and pedestal types, that need to be displayed. How to best display these awards is your decision.
- Separate storage is required for promotional sales gifts, such as logo caps and jackets, as well as company brochures. This same space is to be used as a coat closet and storage for his construction hardhat and steel-toed boots.
- The private bath includes a lavatory, a water closet, and a vanity with a mirror. There is a separate handicap facility in the building.
- Plan the office, bath, and storage space to satisfy the functional, physical, and psychological needs of the client. Create several spatial relationships with the required furniture and equipment, and decide on the best solution for presentation.

Design

- All new finishes, window coverings, lighting, fixtures, and furnishings are required. The design style requested is contemporary, and wood is the preferred material for the case goods. Existing windowsills are white marble (1″ thick and 19″ deep).

Furniture

- A desk with legal file storage, pencil drawer, and box drawers. The minimum size should be 36″ × 72″. Dalton would like to see the actual casework before he purchases it. Contact the manufacturer of the desk that you select to find the nearest showroom that has a display of the desk.
- A return or bridge surface with book storage underneath. Allow for maximum surface area.
- A kneehole credenza with two lateral file drawers and storage capabilities
- An executive, swivel/tilt chair with casters. Dalton has requested leather upholstery.
- Three side chairs with fabric upholstery
- One sofa that can seat three people or three additional side chairs with fabric upholstery
- One occasional or coffee table that is large enough to review blueprints on
- Side tables
- Lamps
- Display for awards
- Artwork and plants

Equipment

- Computer
- Telephone
- Adding machine

SUBMITTAL REQUIREMENTS

Phase II: Conceptual Design   Date Due: _____

A. Develop a preliminary plan based on the functional and aesthetic requirements necessary, including furniture layouts and fixtures.

Evaluation Document: Conceptual Design Evaluation Form

Phase III: Design Development   Date Due: _____

A. Based on the approved preliminary plan, develop and present for review and approval final interior design recommendations to establish and describe the character of the interior space of the project.

B. Prepare documents and illustrations in the form of plans, elevations, color and material sample boards, photographs of furnishings and fixtures, and color rendering.

Evaluation Document: Design Development Evaluation Form

Phase IV: Final Design Presentation   Date Due: _____

A. The final presentation will be an oral presentation to a panel jury. The project must be submitted on 20″ × 30″ illustration boards. Your name, the date, the project name and number, the instructor's name, and the course title must be printed on the back of each board in the lower left-hand corner. Submit as many boards as required, including renderings, elevations, fixture and furniture selections, material selections, and plans.

B. The plans should be presented in 1/4″ scale. The elevations are to be presented in 1/2″ scale.

Evaluation Document: Final Design Presentation Evaluation Form

Phase V: Construction Documentation

Specifications   Date Due: _____

Complete furniture specification forms for all furniture selected. Key the furniture plan to the specification sheets. Insert in a binder. Submit the furniture plan on 11″ × 17″ white bond paper. Submit furniture specifications on 8½″ × 11″ white bond paper and insert all in a black three-ring binder. The binder cover is to be identified in the lower right-hand corner with your name, the date, the project name and number, the instructor's name, and the course title. Sample forms are provided in Chapter 31.

Working Drawings   Date Due: _____

The drawings and specifications must include:

1. A plan indicating any special design elements and attached interior fixtures keyed to details, sections, and elevations as required

2. Elevations or details of walls showing design elements, fixtures, and color and material indications keyed to details and finish specification schedules

3. A floor covering plan and details keyed to specification schedules

4. Details showing sections through specially designed elements or fixtures and identified as to wall locations

5. A lighting criteria plan keyed to a lighting fixture schedule

6. An electrical power criteria plan keyed to a legend

7. A reflected ceiling plan indicating new ceiling conditions, materials, and details with specifications, locating HVAC diffusers as follows: two supply air and one return air

8. Color and material schedules specifying all materials and finishes

9. A post occupancy evaluation questionnaire

Submit the plans on 11″ × 17″ white bond paper and insert in the binder.

Purchase Orders Due Date: _____

Complete purchase orders for the furniture specifications. Complete a client invoice, listing purchase orders and requesting payment for furniture purchases. Both should be on 8½″ × 11″ white bond paper. Insert in the binder. Sample forms are provided in Chapter 31.

Evaluation Document: Construction Documentation Evaluation Form

Project Time Management Schedule: Schedule of Activities

The AutoCAD LT 2002 drawing file name is dalton.dwg and can be found on the CD.

REFERENCES

Books

DiChiara, Joseph, Julius Panero, and Martin Zelnick. 2001. *Time-saver standards for interior design and space planning. 2d ed.* New York: McGraw-Hill.

Harmon, Sharon Koomen, and Katherine E. Kennon. 2001. *The codes guidebook for interiors.* New York: Wiley.

Panero, Julius, and Martin Zelnick. 1999. *Human dimension and interior space: A source book of design reference standards.* New York: Watson Guptill.

Special Resources

Office of Environmental Health and Radiation Safety (EHRS), Ergonomics Assessments: http://www.ehrs.upenn.edu/programs/occupat/furniture.html

Samples of Furniture Specification Sheet, Purchase Order, and Invoice can be found in Chapter 31. You will find an occupant survey in *Professional Practice for Interior Designers, 3d edition,* by Christine M. Piotrowski (New York: Wiley, 2002, p. 551).

Product Manufacturers

Bentley Mills: http://www.bentleyprincestreet.com

Brayton International: http://www.brayton.com

Nienkamper Contract Furnishings, ICF/Group: http://www.nienkamper.com

Pallas Textiles: http://www.pallastextiles.com

NOTES

CHAPTER 3

Conference Room

OUTCOME SUMMARY

The project focuses on exposure to conference/boardroom and presentation room requirements, selecting appropriate contemporary art, and preparing a furniture budget. The project features four phases of the design process: identifying and analyzing the client's needs and goals, developing conceptual skills through schematic or initial design concepts, detailing and refining ideas from the schematic design phase, and drafting documents in preparation for the bidding and contracting of construction, fixtures, and furnishings.

client profile

William A. Wood and Sons, Inc. is a national distributor of metal products, such as steel and aluminum, and has been in business for over 50 years. The products range from rolled metals to perforated and expanded sheet metals, and everything in between. Mr. Mark Dalton, the general manager of the regional distribution center located in Philadelphia, Pennsylvania, has hired you to design the conference space, including the small vestibule outside the room. It is the first phase of a complete remodel of the second floor of the building, which will finally include two additional departments. This is the first major remodel in many years and it is considered visibly necessary. Because of the adjacency of the plant and warehouse facilities to the offices, Dalton has battled with serious soiling issues. The materials and finishes have absorbed years of black stains and soot. Originally built in 1945, the walls are plaster finish over furring strips (excluding the wall separating the conference room from general office areas), and the floor is concrete slab construction. Mr. Dalton has asked for your expertise to help him visualize an interior that is appropriate for Wood's present and future identity. In short, prepare them for the future.

The conference room is used for employee conferences, client conferences, sales meetings, staff meetings, team meetings, teleconferencing, and future video conferencing. Dalton does not want any clutter to be in your head when you walk in the room and has asked you to use the same restraint when designing the interior. "Professional-looking simplicity" is the phrase he used. The selection of materials and the craftsmanship must be of high quality throughout, giving the project an image that is forward-looking and sophisticated.

Fine art and fine commerce are not equally exclusive, and this client is a serious art collector. You are directed to select a work of art that was made after 1950 to be a focal point in this conference room. The client's aesthetic preferences tend to be more contemporary, and the possibilities are endless. You are encouraged to visit contemporary galleries and museums (in person or virtually) to note how contemporary art is displayed. Pay particular attention to the use of nontraditional materials; consider the unique specifications they might require regarding lighting, heating, longevity, as well as related conservation concerns.

After you make your selection, identify the work, the artist, the date, the medium, the size, and the museum collection where it is housed permanently. Determine the artwork's placement in your overall scheme for the conference room while you incorporate it into the project. Include in your projections how the work of art is suited for the interior space, how it responds to the architectural structure itself, and how it connects with the client's industry. Be sure to specify what considerations regarding the environment are necessary (e.g., lighting, humidity). When making your final design presentation, you are to include your rationale for the work of contemporary art along with your furniture and other materials.

project details

General and Architectural Information

Location: William A. Wood and Sons, Inc., Philadelphia, Pennsylvania. Conference room renovation

Square footage: 660

Project budget: Funding for the project has been committed from the corporate capital expenses this year, and the furniture budget allows for an expense of $70,000.00. The furniture, floor covering, lighting, electrical, and new drywall work will be sent out to bid by the owner; therefore, you will be required to submit furniture specifications as part of the construction documents.

Existing construction: The current facility will be renovated. Current facility plans are available. Site information gathering is required.

Consultants required: architect, electrical engineer, videoconference technician, and mechanical engineer

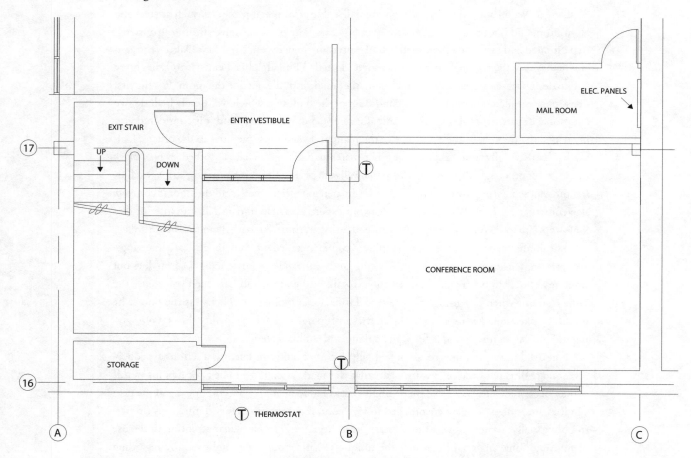

Figure 3–1 Floor Plan

Codes and Governing Regulations

Occupancy: B – Business

Fire suppression: Standpipe and Hose System

Design the project to comply with the most current edition of the BOCA National Building Code (NBC), the National Electrical Code (NEC), the Uniform Fire Code (UFC), and the Americans with Disabilities Act for office spaces.

The owner and contractors will apply for any permits required for the construction work.

Facilities Required

CONFERENCE ROOM

Planning

- Plan a conference table with the ability to seat sixteen, including proper circulation spaces around the table. Allow good sight lines for viewing the media screen.
- One wall will be the presentation wall for a multifunction presentation cabinet and projection screen. Allow for additional circulation space in this area.
- Provide a television, audiovisual, projection, and video equipment area, including storage that is concealed when not in use. Include setup space and equipment for distance projection that is stored when not in use.
- Plan a beverage service area of approximately eight linear feet, with storage for serving items and conference room meeting items and accessories.
- Include a reception or informal meeting area with seating for four. There is no waiting area space in the vestibule outside the conference room.

Design

- All new finishes, trim, window coverings, lighting, fixtures, and furnishings are required. Window covering must have black out capability for proper television, monitor and projection viewing. The design style requested is contemporary, and wood is the preferred material for the case goods. A well-appointed look complete with every detail is required.
- Existing windowsills are white marble (1″ thick and 19″ deep).
- Design a lighting system that includes dimming capabilities with multiple switch locations. Provide good indirect lighting.
- Maximum ceiling height is 11′0″. Minimum ceiling height is 8′5″, which occurs above the window glazing.
- Select neutral, well-appointed elements and materials for the vestibule that will help make the transition to the adjacent departments.

Furniture

- A conference table with electrical hookup capabilities for slide projector, overhead projector, and laptop computer or monitors. Allow three square feet per person at the table.
- Large comfortable swivel/tilt leather chairs with carpet casters for conference seating
- Wall presentation cabinet that opens to a total width of eight linear feet, including a dry marker board, tack surfaces, and paper pads
- Two floor easels
- An easel ledge trim designed to attach to the wall on either side of the presentation cabinet
- Projection screen with a minimum width of 8′0″
- Audiovisual carts for projection equipment that are to be stored when not in use
- Storage and serving furniture components, including drawers, wardrobe for coat storage, and shelving

- Sofa or chair arrangement with related furnishings for the reception area
- Complete conference accessories, including individual table pads, portfolios, pencil cups, tablet holders, coasters, serving trays, water pitchers, waste cans, and so on
- Plants, accessories, and artwork

Equipment

- Computer and laptop LAN connections
- One wall-mounted telephone and one movable phone for the conference area and the seating group
- Television, overhead and slide projector, VCR and DVD equipment, and film projector
- Future video and teleconferencing equipment including two cameras, two microphones, two monitors, and a portable control panel for an Integrated Services Digital Network (ISDN)
- Fire extinguisher

SUBMITTAL REQUIREMENTS

Phase I: Programming   Part A Due: _____ Part B Due: _____

A. Research various codes, arrangements, and clearances for the above-outlined spaces. Research fixtures, furniture, and equipment for the same.

B. Provide a one-page typed concept statement of your proposed planning direction and initial design concept.

Presentation formats for Parts A and B: White bond paper, 8½″ × 11″, with a border. Insert in the binder or folder.

Evaluation Document: Programming Evaluation Form

Phase II: Conceptual Design Part A Due: _____ Part B Due: _____

A. Develop a preliminary partition and furniture plan, showing all walls, ceilings (with suggested locations of HVAC linear air diffusers), furniture, built-in fixtures, doors, and windows. Develop a preliminary lighting plan and power plan. These plans can be freehand drawings or CADD drawings in 1/4″ scale. Drawings must have an appropriate border. Finished drawings should be a blueprint or white bond copy.

B. Develop the above plans into a final floor plan, showing all items noted in the program. This plan is to be completed using the CADD program. Develop ceiling and lighting studies into a final reflected ceiling plan, showing all ceiling conditions and heights using the CADD program.

Evaluation Document: Conceptual Design Evaluation Form

Phase III: Design Development   Part A Due: _____ Part B Due: _____

A. Develop a color, material, and final furniture scheme. Samples and pictures can be presented in a loose format.

B. Develop a perspective sketch of the interior of the room, showing all design elements and furnishings.

Evaluation Document: Design Development Evaluation Form

Phase IV: Final Design Presentation   Date Due: _____

A. The final presentation will be an oral presentation to a panel jury. Project must be submitted on 20″ × 30″ illustration boards. Your name, the date, the project name and number, the instructor's name, and the course title must be printed on the back of each board in the lower left-hand corner. Submit as many boards as required, including a colored rendering, fixture and furniture selections, material selections, floor plan, ceiling plan, and power plan.

B. The plans should be presented in 1/4″ scale. The elevations are to be presented in 1/2″ scale.

C. Complete a furniture budget and indicate lead times required for items specified. Present on 8½″ × 11″ white bond paper. Insert in the binder.

Evaluation Document: Final Design Presentation Evaluation Form

Phase V: Construction Documentation   Date Due: _____

A. Prepare working drawings and specifications for non-load-bearing interior construction, materials, finishes, furnishings, fixtures, and equipment for client's approval. The plan drawings presented in 1/4″ scale, and the elevations presented in 1/2″ scale. The drawings and specifications must include:

1. A plan indicating any special design elements and attached interior fixtures keyed to details, sections, and elevations as required
2. A furniture plan keyed to furniture specifications.
3. Elevations and details of walls showing design elements, fixtures and color and material indications keyed to details, and finish specification schedules
4. A floor covering plan and details keyed to specification schedules
5. Details showing sections through specially designed elements, or fixtures identified as to wall locations
6. A lighting criteria plan keyed to a lighting fixture schedule
7. An electrical power criteria plan keyed to a legend
8. A reflected ceiling plan indicating new ceiling conditions, materials, and details with specifications, and a suggestion for the location of linear air diffusers in the ceiling for HVAC
9. All materials and finishes specified on color and material schedules
10. Title sheet, including index of drawings, project representatives, applicable governing codes, project statistics, utility service representatives, and title block
11. A floor covering maintenance plan and furniture maintenance recommendations

Evaluation Document: Construction Documentation Evaluation Form

Project Time Management Schedule: Schedule of Activities

The AutoCAD LT 2002 drawing file name is conference.dwg and can be found on the CD.

REFERENCES

Books

Ballast, David Kent. 1994. *Interior construction and detailing for designers and architects.* Belmont, Calif.: Professional.

DiChiara, Joseph, Julius Panero, and Martin Zelnick. 2001. *Time-saver standards for interior design and space planning. 2d ed.* New York: McGraw-Hill.

Harmon, Sharon Koomen, and Katherine E. Kennon. 2001. *The codes guidebook for interiors.* New York: Wiley.

Panero, Julius, and Martin Zelnick. 1999. Human dimension and interior space: A source book of design reference standards. New York: Watson Guptill.

Product Manufacturers

Brayton International: http://www.brayton.com

Davis Furniture Industries, Inc.: http://www.davis-furniture.com

Herman Miller, Inc.: http://www.hermanmiller.com

Knoll, Inc: http://www.knoll.com

Maharam: http://www.maharam.com

Mohawk International: http://www.mohawkgroup.com

Nienkamper Contract Furnishings, ICF/Group: http://www.nienkamper.com

Tella: http://www.tellainc.com

NOTES

CHAPTER 4

PSR

OUTCOME SUMMARY

The project focuses on exposure to systems furniture and office planning. The project features four phases of the design process: identifying and analyzing the client's needs and goals; developing conceptual skills through schematic concepts; detailing ideas from the schematic design phase; and specification documents, including plans and furniture specifications for the bidding and contracting of system furnishings and their interface with the building structure.

client profile

PSR Associates, the developer, has hired you to develop an office systems furniture specification and budget estimate for tenant lease spaces #201 and #301 at Pointe Plaza. John Anthony is your client contact, and his plan is to include a furniture specification that will be an optional part of the tenant lease allowance. Using a predetermined allowance of space allocation, Anthony wants you to develop several typical open plan layouts using systems furniture, and the specifications for the furniture. The plans are to incorporate several typical plans, including Team Work Group arrangements and Free-address offices. The typical requirements are outlined below. It is important that the plan and furniture solution for each station considers the mechanical interface of the systems furniture with the building service. A system with integrated raceways to carry system cables is essential for each station. The building is equipped with fiber optic cables and you are to assume that the tenants' telephone and data will be on a local area network (LAN) system.

You are to incorporate the following considerations when designing:

- Sufficient privacy and quiet, within the open space concept. Surface materials and component construction to improve acoustics
- Ample working space, in plan, and for work surfaces and storage
- High-quality indirect lighting
- Comfortable, adjustable seating
- Space that promotes the ability to personalize the workstation
- State-of-the-art equipment and furniture

Pointe Plaza is an office building located on a championship golf course just minutes away from the golf club, hotel, and a proposed conference center. Additionally, the site is 20 minutes from the downtown metropolitan area and the international airport. Building services include monitored security systems, fiber optics, and a perk of preferred membership privileges at the golf club. PSR offers full-service, five-year leases. The interior is finished with ecru-colored modular carpet tiles, textile wall covering, and acoustical tile ceilings. Sound control has been provided through masking with a fixed-volume ventilation system.

Figure 4-1
Rendering of Pointe Plaza

project details

General and Architectural Information

Location: Pointe Plaza at Southpointe, 601 Technology Drive. Lease space #201, second level; and #301, third level

Developer: PSR Associates

Building area: 41,200 gross square feet. The building has three stories. Average floor area: 13,000 rental square feet

Square footage of project: 7,500 (net) second level; 7,000 (net) third level

Project budget: Minimal cost with maximum flexibility

New construction: Current facility plans are available.

Consultants required: Architect, electrical engineer, and mechanical engineer

Codes and Governing Regulations

Occupancy: B – Business

Fire Detection and Suppression: Automatic Sprinkler System (wet and dry)

Design the project to comply with the most current edition of the International Building Code (IBC) and the Americans with Disabilities Act for office spaces.

Facilities Required

- Receptionist Station (1), Executive Assistant Station (6), Mid-Manager and Customer Service Stations (40), Team Work Groups (12), and Free-Address Station (10)

RECEPTIONIST STATION

- Approximately 96 square feet. One station is required.
- Panels must be a combination of desk height and counter height.
- Plan a transaction counter in front.
- Provide space for a telephone, a computer, a fax machine, a printer, and a small typewriter.
- Include at least 6 linear feet of lateral files.

EXECUTIVE ASSISTANT STATION

- Approximately 48 to 96 square feet. Six stations are required.
- Panels must be of desk height.
- Work surfaces must include a return for the computer or typewriter.

MID-MANAGER STATION AND CUSTOMER SERVICE STATION

- Approximately 125 to 150 square feet. Forty are required.
- Panels must provide stand-up privacy.
- Plan flexible open workstations, and locate twelve managers adjacent to each of the 12 Team Work Groups.

TEAM WORK GROUPS

- Approximately 100 square feet per member. Team of five members: seven are required. Team of three members: five are required.
- The five-member team space and the three-member team space must include individual work areas and a common meeting space for each team.
- Plan for a team meeting area to assemble for project discussions that includes a table and chairs for at least each team member.
- Panels must provide sit-down privacy.

FREE-ADDRESS STATION

- Approximately 100 square feet. Ten are required.
- The ten unassigned offices must be identical.
- Panels must provide sit-down privacy.

Planning

- Aisle circulation space must be a minimum of 44″ in width.
- Allow for flexibility and creativity. Plan to support and reinforce communications, and locate twelve managers adjacent to each of the 12 Team Work Groups to enhance and encourage the necessary interactions.
- Typical workstations must include a maximum amount of work surface area, computer space, file storage, storage for reference materials and office supplies, and work tools. Locate task seating "with no backs to the doors" or openings into work areas wherever possible.
- Research and use a predetermined allowance of space allocation for job level functions. Plan each one of the above workstations using systems furniture.

Design

- Surface materials and component construction must be selected to improve acoustics. Panels must have a sound transmission class (STC) rating over 23.
- Incorporate the use of fluid shapes, which minimize the need for excessive lateral movement when working.

Furniture

- Panels and components, including work surfaces, shelves, drawer units, flipper doors, EDP shelves, file units, task lights, tack surfaces, pencil drawers, box drawers, file pedestals and box pedestals or their equal
- Accessories, including paper handling equipment, shelf dividers, paper trays, and keyboard trays
- Adjustable height work surfaces for ergonometric issues
- Fully adjustable task seating, with carpet casters at the workstations and meeting areas

Equipment

- Telephone and headset
- Computer monitor, printer, modem, and cable connections
- Power-operated tools such as radios, pencil sharpeners, electric erasers, and others (assuming their use at the stations)

SUBMITTAL REQUIREMENTS

Phase II: Conceptual Design   Date Due: _____

A. Develop a preliminary plans based on the functional and aesthetic requirements necessary, including use of space, furniture layouts, and fixtures for each of the following workstations:

1. Receptionist Station (one plan)
2. Executive Assistant Station (one plan)
3. Mid-Manager Station (one plan)
4. Team Work Groups (one plan)
5. Free-Address Station (one plan)

Evaluation Document: Conceptual Design Evaluation Form

Phase III: Design Development   Date Due: _____

A. Based on the approved preliminary plans, develop and present for review and approval final interior design recommendations for the systems furniture for each station.

B. Prepare documents and illustrations in the form of plans, elevations, color and material sample boards, and photographs of furnishings.

Evaluation Document: Design Development Evaluation Form

Phase IV: Final Design Presentation   Date Due: _____

A. Prepare a preliminary budget, including cost of each station with the quantities requested. Be prepared to discuss the features of the furniture system chosen, including the psychological, physical, and cost benefits.

B. The final presentation will be an oral presentation to a panel jury. Project must be submitted on 20″ × 30″ illustration boards. Your name, the date, the project name and number, the instructor's name, and course title must be printed on the back of each board in the lower left-hand corner. Submit as many boards as required, including 3D drawings, elevations, furniture selections, material selections, and plans.

C. The plans should be presented in 1/4″ scale. The elevations are to be presented in 1/2″ scale.

Evaluation Document: Final Design Presentation Evaluation Form

Phase V: Construction Documentation   Date Due: _____

The drawings and specifications must include:
1. Plans keyed to details, sections, and elevations as required
2. Elevations or details of furniture showing component elements, and color and material indications keyed to finish specifications
3. Lighting criteria plans for the systems furniture keyed to specifications
4. An electrical power criteria plan for the systems furniture keyed to a legend
5. Complete furniture specification forms for all furniture selected. Key a furniture plan to specification sheets. Insert in the binder. Submit the furniture plans on 11″ × 17″ white bond paper. Submit furniture specifications on 8½″ × 11″ white bond paper, and insert all in a black three-ring binder. The binder cover is to be identified in the lower right-hand corner with your name, the date, the project name and number, the instructor's name, and the course title.

Submit the plans on 11″ × 17″ white bond paper and insert in the binder.

Evaluation Document: Construction Documentation Evaluation Form

Project Time Management Schedule: Schedule of Activities

Use the AutoCAD LT 2002 drawing file template named psr.dwg on the CD. It is a blank drawing template.

REFERENCES

Books

Henderson, Justin. 2000. *Workplaces and workspaces: Office designs that work.* Gloucester, MA: Rockport Publishing, Inc.

Hershberger, Robert. 1999. *Architectural programming and predesign manager.* New York: McGraw-Hill Professional.

Marmot, Alexi, and Joanna Eley. 2000. *Office space planning: Designs for tomorrow's workplace.* New York: McGraw-Hill Professional.

Piotrowski, Christine M., and Elizabeth A. Rogers. 1998. *Designing commercial interiors.* New York: Wiley.

Special Resources

Castelli 3D Design Tool (Computer Software Symbol Library) by Haworth, Inc.:

20-20 Giza, Inc.: http://www.giza.com

Sample of Furniture Specification Sheet can be found in Chapter 31.

Product Manufacturers

Design Options (Information and Price Guide brochure contains typical workstation plans.): http://www.designoptions.com

Herman Miller (CADD packages are available for plan layouts.): http://www.hermanmiller.com

Herman Miller: http://www.eamesoffice.com

Teknion: http://www.teknion.com

USM Haller Modular Furniture System: http://www.usm.com

NOTES

CHAPTER 5

MCA Regional Headquarters

OUTCOME SUMMARY

As a programming problem the project features two phases of the design process: identifying and analyzing the client's needs and goals and developing conceptual skills through schematic concepts. This project brings into focus how the design process may interface with the client's preliminary decision-making process to ascertain the feasibility of proceeding with a proposed expansion. It is a Charette that will provide experience in working within a short four-hour time limit and includes skill development in program analysis and planning.

client profile

Mechanical Contractors Association of America (MCAA) is committed to providing the tools to create a better future for contractors and their business. It is both their vision and mission. The Mechanical Contractors Association's (MCA) regional headquarters serves the northeastern area firms engaged in the installation of heating, ventilation, air-conditioning, and process piping. The association provides services to its members, including labor relations, negotiation and communications, educational seminars and programs, government affairs, engineering services, updates on new technology and regulations, and business services. However, the most important contribution that has been provided by the MCA is the many accord agreements forged between industry labor and management. As a bargaining agent for employees with local trade unions, the MCA has facilitated many peaceful adjustments of jurisdiction disputes in the building and construction industry. MCAA has a long history of peace among the trades, preventing strikes and lockouts and facilitating adjustment of jurisdiction disputes. MCAA has been equally committed to the effort for workers' compensation reform.

The present scenario is that the MCA has outgrown its current facility of 1,240 square feet and has plans to expand on the existing site and increase the space by 1,950 square feet. The current facility lacks space for board meetings, member meetings and negotiating facilities, and support spaces such as a kitchen and a mail and copy room. The MCA would like some planning ideas from you on the use of the proposed expanded 3,190 square feet facility.

The walls are gypsum board over metal studs. The ceilings are 9′0″ above finished floor (A.F.F.) and are 2×2 acoustical tiles. The existing boardroom ceiling is a linear metal (aluminum) suspended ceiling system running lengthwise in the room. You may relocate interior doors and add walls as required. Existing wall partitions must remain.

project details

General and Architectural Information

Location: MCA Regional Headquarters, Northeastern Chapter, 521 Cypress Street

Existing square footage: 1,240

Expansion square footage: 1,950. Gross: 3,190

Initial number of employees in facility: Two

Hours of operation: Generally Monday–Friday 9:00 a.m. to 5:00 p.m.

Existing construction and planned expansion: Current facility plans and specifications are available. Site information gathering is required.

Consultants required: Architect, electrical engineer, and mechanical engineer

Codes and Governing Regulations

Occupancy: Mixed use. B – Business; A2 – Assembly

Fire suppression: Standpipe and Hose System

Design the project to comply with the most current edition of the BOCA National Building Code (NBC), the National Electrical Code (NEC), the Uniform Fire Code (UFC), and the Americans with Disabilities Act.

Facilities Required

Existing space: Lobby, Coat Closet, Ladies Room, and Men's Room will remain in existing locations. Relocate the Director's and Secretary's Offices, relocate the File/Storage Room, create a Mail/Copy Room, create a Pay Phone Area, create an Electrical/Telephone Panel Room, and create an Entry Vestibule. Relocate the kitchen to the new addition. Relocate mechanical equipment to the roof.

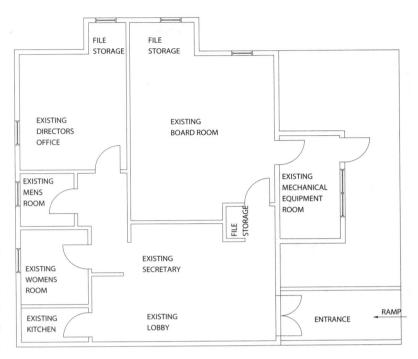

Figure 5–1 Existing Plan

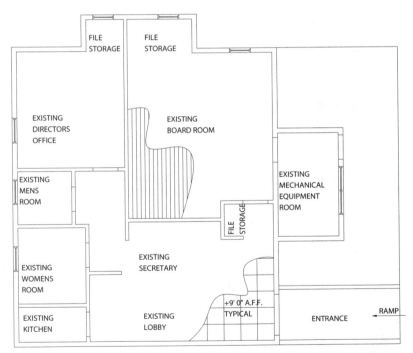

Figure 5–2 Existing Ceiling

New additions: Atrium/Pre-function Area, Meeting Room, Boardroom, Caucus Room, Catering Kitchen, Service Entrance, Table and Chair Storage, and Emergency Exit

Spatial Requirements

DIRECTOR'S OFFICE

Planning
- Adjacency: Secretary's Office. Approximately 200 square feet

SECRETARY'S OFFICE

Planning
- Adjacency: Lobby, Director's Office (Secondary: File/Storage Room and Mail/Copy Room). Approximately 150 square feet

FILE/STORAGE ROOM

Planning
- Adjacency: Secretary's Office. Approximately 200 square feet

MAIL/COPY ROOM

Planning
- Adjacency: Secretary's Office. Approximately 90 square feet

PAY PHONE AREA

Planning
- Adjacency: Lobby

ELECTRICAL/TELEPHONE PANEL ROOM

Planning
- Approximately 30 square feet

ENTRY VESTIBULE

Planning
- Adjacency: Exterior entrance, Lobby. Approximately 30 square feet

LOBBY

Planning
- Adjacency: Entrance, Secretary's Office, Pre-function Room, Ladies Room, and Men's Room. Approximately 225 square feet
- Coat Closet: 4 linear feet

LADIES ROOM AND MEN'S ROOM

Planning
- Adjacency: To remain in the existing location

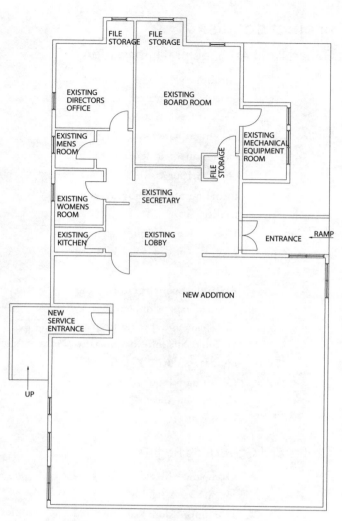

Figure 5–3   Plan of Proposed Addition

ATRIUM/PRE-FUNCTION AREA

Planning
- Adjacency: Lobby, Meeting Room, and Boardroom. Approximately 250 square feet

MEETING ROOM

Planning
- Adjacency: Pre-function Area. Approximately 800 square feet
- Serving area: 10 linear feet
- To seat approximately fifty-five people, classroom style

BOARDROOM

Planning
- Adjacency: Pre-function Room, Caucus Room. Approximately 425 square feet. This room is the nerve center of the organization.

CAUCUS ROOM

Planning
- Adjacency: Boardroom. Approximately 200 square feet. Access will be from the Boardroom only.

CATERING KITCHEN

Planning
- Adjacency: Meeting Room, Service Entrance. Approximately 80 square feet

SERVICE ENTRANCE

Planning
- Adjacency: Catering Kitchen

TABLE AND CHAIR STORAGE ROOM

Planning
- Adjacency: Meeting Room. Approximately 90 square feet

SUBMITTAL REQUIREMENTS

Phase I: Programming

A. Analyze the program outlined above and develop a programming matrix keyed for adjacency locations. Use the sample blank matrix provided in Figure 5–4.

B. With the information gathered from the matrix, provide a minimum of three bubble diagrams for the spatial relationships of the areas required.

Presentation formats for Parts A and B: Tissue overlays
Evaluation Document: Programming Evaluation Form

Phase II: Conceptual Design

A. Develop a preliminary partition and furniture plan, showing all walls, ceilings, built-in fixtures, doors, and windows on the plan provided in Figure 5–5. This plan can be freehand drawings in 1/8″ scale.

B. At the end of the 4-hour time limit, present your solution to the class. Your verbal presentation must include justification for your spatial and functional solutions.

Evaluation Document: Conceptual Design Evaluation Form

The AutoCAD LT 2002 drawing file name is mca.dwg and can be found on the CD.

MATRIX

Lobby
Coat Closet
Ladies Room
Men's Room
Director
Secretary
File /Storage Room
Mail/Copy Room
Pay Phone Area
Panel Room
Entry Vestibule
Catering Kitchen
Atrium/Pre-function Area
Meeting Room
Board Room
Caucus Room
Service Entrance
Storage
Emergency Exit

LEVEL OF INTERFACE, COMMUNICATION OR ADJACENCY

● PRIMARY
○ SECONDARY
◇ MINIMUM TO NONE
— UNDESIRABLE

Figure 5–4   Matrix

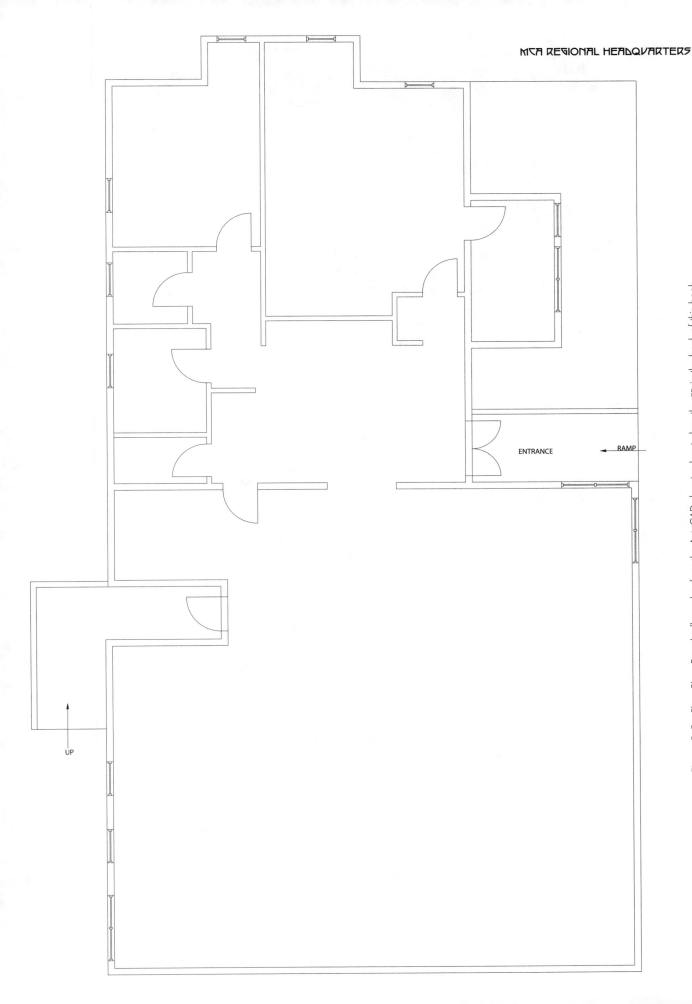

UP

ENTRANCE

RAMP

Figure 5–5 Floor Plan. Print the floor plan from the AutoCAD drawing located on the CD in the back of this book.

CHAPTER 6

LiteLine Products

OUTCOME SUMMARY

The project is a space-planning problem for three small divisions within a larger corporation, including the identification and analysis of the client's needs and goals. It will concentrate on the programming phase of the design process, focusing on skill development of project management in organization of information, analysis, spatial organization, and adjacency requirements.

client profile

LiteLine is the stainless and aluminum metal division of William A. Wood and Sons, Inc., a national distributor of metal products such as steel and aluminum. Mr. Mark Dalton, the general manager of the regional distribution center located in Philadelphia, Pennsylvania, has hired you to do some preliminary programming and planning on the second level for the LiteLine product division, the Programming Department and an office space for the national accounts manager. Tom Silvers is the manager of LiteLine; Joe Manetti is the chief programmer, and Earl Benny is the national accounts manager.

Dalton, Silvers, and Manetti have given great consideration to the current issues of their workspaces. The deficiencies of this work environment are workspace, privacy, quiet, storage space, work surfaces, security, and individual wall space. The equipment is poorly organized with no accommodations made for the many wires associated with the computers and peripherals.

They did some brainstorming within the work groups and have asked you to help them develop a space that is intuitive, environmental, light, and flexible. Tom further characterized their desire for a relaxed, respectful, harmonious, and advantageous arrangement for teamwork. Each employee requires a personal workspace that is generous and comfortable, with good storage and which addresses physical ease. Silvers and Dalton have asked for adequate privacy and lower noise levels within the open-space concept and adequate working space in general. Proper lighting and ergonomic seating are big requests from the employees. They would also like a little wall space to promote privacy and the ability to personalize it by hanging pictures and other items.

The building was originally built in 1945. The walls and ceilings are plaster finish over furring strips, and the floor is concrete slab construction. You may core drill to locate new power or communications outlets. The window elevations and details are the same as those found in Dalton's office, which has white marble windowsills (1″ thick and 19″ deep).

project details

General and Architectural Information

Location: William A. Wood and Sons, Inc., Philadelphia, Pennsylvania. Second-level office renovations of LiteLine, Programming Department, and office of National Accounts manager

Gross square footage: 1,825

Initial number of employees in facility: 11. Projected number of employees: 13

Project budget: Funding for the project has not been committed from the corporate capital expenses this year.

Existing construction: The current facility will be renovated. Current facility plans are available. Site information gathering is required.

Consultants required: Architect, electrical engineer, and mechanical engineer

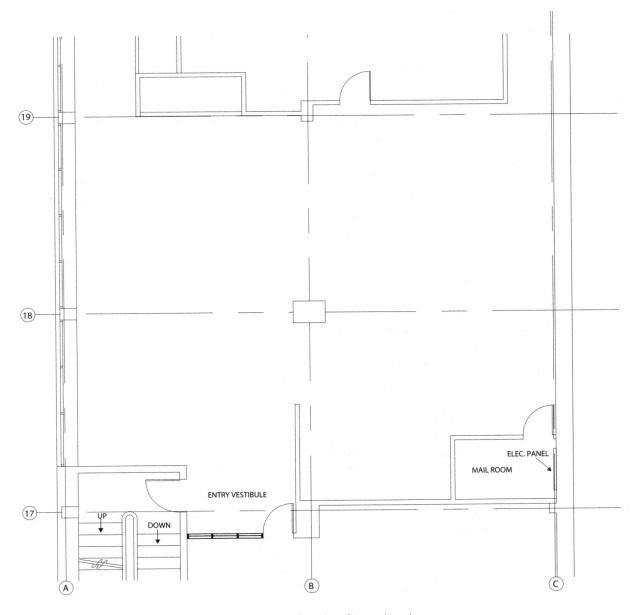

Figure 6–1   Floor Plan of Second Level

Codes and Governing Regulations

Occupancy: B – Business

Fire suppression: Standpipe and Hose System

Design the project to comply with the most current edition of the BOCA National Building Code (NBC), the National Electrical Code (NEC), the Uniform Fire Code (UFC), and the Americans with Disabilities Act for office spaces.

Facilities Required

- LiteLine Products, Programming Department, Office of National Accounts Manager

LITELINE PRODUCTS DIVISION

General Planning Information

As the general manager of LiteLine, Tom oversees the plant operation, has marketing responsibility for the entire product line, and has purchasing responsibility. The department requires four stations and an additional station if space allows. Without a doubt, this department functions as a team. Working closely together, the department employees have long workdays, averaging 10 to 11 hours. In general the employees need more desktop workspace, more overhead shelf storage for books and manuals at each station, and more file storage. Storage for office supplies in a central location (in a credenza now) has also been requested. A central area for files, copier, and fax is required, and the copy machine will be shared with all second-floor departments. The central fax will be shared with the national account manager.

LiteLine General Manager

- Tom spends approximately 5 hours a day on the telephone and 3 hours on the computer. Most office personnel communications is primarily with the product specialist, Debbie, and with Ed, inventory and quality claims service specialist. Most documents handled are letter size. Tom prefers a window location, privacy, and free movement.
- Tom would like a desk-height work surface a minimum of 72″ × 36″, credenza surfaces a minimum of 72″ × 20″, and a bridge computer surface with retractable keyboard tray a minimum of 30″ × 20″. He has book and manual storage requirements of 12 linear feet of shelving. He prefers an executive chair, two guest chairs, and a side table.

LiteLine Senior Product Specialist

- Debbie communicates very frequently with the plant and with other department members via E-mail or directly as the senior product specialist of LiteLine. She is on the computer nearly constantly, on the telephone about 50% of the day, and at the copy machine for approximately 20 minutes a day.
- She works with letter size, legal size, and 11″ × 17″ paper documents. She requires 90 linear inches of shelf storage and 150 linear inches of letter file storage. There is a need to access to contracts, which are remote shared files.
- A large work surface for spreading out files and situating the computer and guest chair at her station is required.

LiteLine Merchandise Assistant

- Jason, the merchandise assistant, prefers to maintain close proximity to the entire team because he communicates several times a day with all LiteLine employees. He spends 3 hours on the telephone per day and uses the printer, copier, and fax several times a day.
- Working with letter-size documents, he requires 100 linear inches of file storage, storage for stationery and envelopes, and remote storage for disks and other files. He needs 100 linear inches of shelf storage.

LiteLine Inventory and Quality Claims Service Specialist

- Ed Walls is responsible for quality claims and local inventory issues and requires a space free from visual distractions for concentration. He has constant contact with the senior product specialist, Debbie, Sherri, the quality claims assistant, Jason, and Dan from the warehouse. Good communication access to each of these people is required. Ed spends about 20 minutes of the day in communication with Tom, and, he spends 3 hours on the telephone. Ed is almost constantly on the computer and calculator, and uses the copy machine approximately five times a day.
- Ed works with reference books, files, and letter-size documents and requires 96 inches of book storage and 72 linear inches of letter storage. In addition, he needs an area to display Quality Assurance (Q.A.) charts and results, and his memo reminders.
- Provide one guest chair.

LiteLine Quality Claims Assistant

- Sherri works closely with Ed as an assistant and requires a typical clerical workstation with a computer and a telephone.
- Two letter-size file drawers, stationery storage, and two box drawers are required.
- General shared files and required storage are: 30 linear feet, legal; 9 linear feet, binder storage; and 3 linear feet, disk storage.

PROGRAMMING DEPARTMENT

- Programming is part of W.W. Wood and includes four programming stations and one clerical station. The programmers are responsible for the sophisticated computerized distribution tracking system and special or custom production orders. Programming is a highly stressful job. Efficiency is required working on the computer, fax, line printer, and telephone. Currently, the work surface areas are inadequate. The company has requested that each programmer have a minimum of 18 square feet of work surface area and four electrical receptacles at each station.
- Most documents handled are letter size, and the line printer and the fax are used quite frequently. Each programmer requires 100 linear inches of file storage, storage for stationery and envelopes, and remote storage for disks and other files.
- If space allows, plan for an additional programming workstation.
- General file storage requirements are: 12 linear feet, legal files; 100 linear feet, letter files; and 12 linear feet, disk storage. This file storage should be adjacent to the programming stations

NATIONAL ACCOUNTS MANAGER

- Earl is the national accounts manager handling ten multilocation national accounts and communicates frequently with LiteLine and all of the W.W. Wood plants to coordinate the marketing efforts. Workdays are normally 9½ hours. He spends approximately 2 to 3 hours on the telephone per day. About 60% of his time is spent in the office, and the balance is spent traveling for sales and client meetings. Earl prefers a window location and large work surfaces.
- He works with letter-size documents and uses the copy and fax machines daily. A printer is used only two to three times a week. Working mainly on his laptop computer, he would like a dedicated line to the mainframe computer.
- Earl has requested 180 linear inches of shelf storage, 324 linear inches of letter file storage, and 36 linear inches of legal file storage. Storage is also required for business forms and stationery. Other needs are secured spaces for the laptop computer and disk storage.
- Provide a display area for a large map of the United States locating national accounts, and space for guest seating in his office.

- Because he spends 30% of his time on the telephone, he requires a multiple-line telephone with special features, including VMX, call forwarding, auto call back, hands-free dialing, message box, redial, abbreviated dialing, recall, hold, transfer, conferencing, and speaker.

SUBMITTAL REQUIREMENTS

Phase I: Programming
Part A Due: _____   Part B Due: _____   Part C Due: _____

A. Analyze the information presented in the client profile and details, and prepare an outlined program for each department and personnel, using the following outline or one with a similar objective:

1. (Name of Department and personnel and job title)
   a. Planning
      i.   Adjacency
      ii. Square Footage
   b. Design
   c. Furniture
   d. Equipment

B. Following your program outline, develop a programming matrix keyed for adjacency locations. A sample blank matrix is provided in Chapter 31.

C. With the information gathered from the matrix, provide a minimum of three bubble diagrams for the spatial relationships of the areas required.

Evaluation Document: Programming Evaluation Form

Phase II: Conceptual Design
Part A Due: _____   Part B Due: _____

A. Develop a preliminary partition and furniture plan, showing all walls, ceilings, built-in fixtures, doors, and windows. These plans can be freehand drawings or CADD drawings in 1/4″ scale. Drawings must have an appropriate border. Finished drawings must be a blueprint or white bond copy.

B. Develop the above plans into a final floor plan, showing all items noted in the program. This plan is to be completed using the CADD program. Develop ceiling and lighting studies into a final reflected ceiling plan, showing all ceiling conditions and heights on the CADD program.

C. Your name, the date, the project name and number, the instructor's name, and the course title must be printed on each drawing. Submit as many drawings as required, including plans. The plans should be presented in 1/4″ scale.

Evaluation Document: Conceptual Design Evaluation Form

The AutoCAD LT 2002 drawing file name is liteline.dwg and can be found on the CD.

REFERENCES

Refer to previous office projects.

NOTES

CHAPTER 7
Danna Industries

OUTCOME SUMMARY

The project features a team approach to three phases of the design process: to define design team responsibilities, discuss goals, and explore client requirements through design team meetings, buzz sessions, and dialogue groups; to establish optimal space plans using adjacency diagrams, which show the proximity of the various work groups; and to refine the design concept conveyed through drawings, sketches, material samples, and space plans for a final design presentation. The team is responsible for researching the local codes and arranging a mock preliminary plan check meeting with local code officials to review life safety and accessibility for office spaces.

client profile

The client, Mark Danna, of Danna Industries (DI), is both a personal and professional associate of yours. Since 1946 the "Total Solution" mechanical construction firm has been serving the needs of the industry and its clients, including mechanical installation, sheet metal construction, and service. The types of building specializations of DI are large retail facilities, educational institutions, hospitals and medical facilities, correctional institutions, commercial office buildings, sports complexes, airports, country clubs, theaters, hotels, conference centers, and performing art centers. The mission statement of Danna Industries is "To be the best of the top three Mechanical Contractors that clients prefer to have build and service their HVAC and Plumbing projects. We strive to be distinctive from the others by pushing our-

Figure 7–1   DI Project: Adaptive reuse of the Allegheny County Jail, originally designed by Henry Hobson Richardson. Now used as a juvenile courthouse.

selves to deliver superior service through knowledge and exceptional, on time, done right the first time, delivery." As a team, DI's employees partner with their clients to build projects that achieve tangible results. Key elements of their goals include fair pricing, project planning, multitrade coordination, and, most importantly, "On-Time Construction."

Complete with modernized equipment, the sheet metal plant will meet any demand. With floor space of over 30,000 square feet, DI has designed the plant for the highest efficiency, including full CAD/CAM project integration from layout to fabrication. Several million pounds of product leave the plant each year on a single work shift. Quality is never sacrificed, and DI believes in the old saying, "build it right the first time." Danna Industries has invested over one million dollars in developing one of the most innovative state-of-the-art plants, making it a full-service mechanical contractor who does not depend on a subcontractor to perform sheet metal work. This investment increases productivity, saves time, reduces costs, and enables DI to meet any aggressive schedule to get the job done.

DI believes that its employees are its most valued asset, and invests in all employees to help them reach his or her full potential, which ultimately benefits its customers. Recognizing that people have individual needs and desires, DI continually invests in education and training to enhance the staff's personal and professional development. "This commitment to our staff promotes the T. E. A. M. mentality that Danna Industries strives to maintain. Every employee has a direct and tangible effect on the outcome of every project and it is in working Together that our Efforts help us to Achieve our Mission."

Danna Industries has recently purchased a 70,000 square foot facility to house the administrative offices, plant, and warehouse. Situated on a hill with an abundance of land and foliage, views from the building are magnificent. Danna would like to bring the same feeling to the interior with plants and what is sometimes called "green wall." The structure is brick and concrete block construction with concrete slab floors. The existing ceiling is a 2×4 suspended grid on the first level. The mezzanine has an exposed truss ceiling.

Teamed with Tony, Joe, and Mark Danna, you are responsible for planning and designing the administration facilities. The solution should create an image that reflects and interprets the firm's attitudes as described earlier. The plan must be of a well-organized functional arrangement that encourages productivity. The key planning issues are communications, work group interaction, and job adjacencies. Incorporate the use of conventional office furniture and office systems furniture in your solution. Use conventional full-height wall partitions and window locations for the president, the vice presidents, and the project managers. Skylights are being added for the other staff's offices. Provide a uniform ambient level of illumination, adequate task light levels in work areas, and accent lighting. Lighting design solution must also reduce glare and eyestrain.

Include an emergency station quadrant with a designated evacuation for every ten employees. Each quadrant must have a designated exit plan and a storage location for related supplies, including a battery-powered radio, a blanket, a flashlight, a two-way radio, and a first aid kit. At least two employees from each quadrant have taken a first aid and AED/CPR course.

Figure 7–2
Sheetmetal ductwork installation
(Courtesy: Brud Bavera, Photographer.)

Figure 7–3
Equipment installation

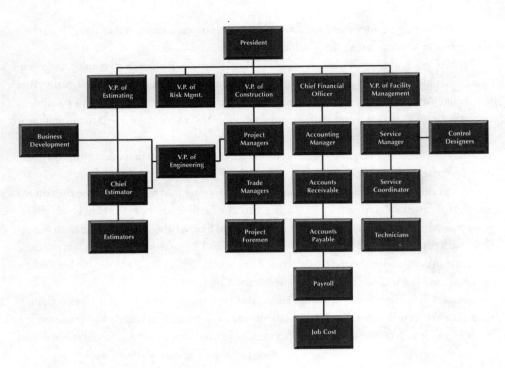

Figure 7–4 Organizational Chart

project details

General and Architectural Information

Location: Danna Industries, Administrative Offices, 381 Manor Road

Gross square footage: 16,475. First level: 14,000 square feet; Mezzanine: 2,475 square feet

Project budget: The furniture, floor covering, lighting, electrical, and new drywall work will be sent out to bid by the owner; therefore, you will be required to submit furniture specifications as part of the construction documents.

Existing construction: The existing facility will be renovated. Current facility plans are available. Site information gathering is required.

Consultants required: Architect, electrical engineer, and mechanical engineer

Codes and Governing Regulations

Occupancy: Mixed use. I – Industrial; B – Business; A2 – Assembly

Fire suppression: Standpipe and Hose System

Design the project to comply with the most current edition of your local adopted building codes and the Americans with Disabilities Act for office spaces.

The owner and contractors will apply for any permits required for the construction work.

Facilities Required

First Level: Entry Vestibule and Reception, President's Office, Executive Administrator, Main Conference Room, Vice President of Facility Management Office Administrative Assistant, Vice President of Construction Office, Administrative Assistant of Construction, Project Managers' Offices (4), Assistant Project Managers (2), Administrative Assistant to Project Managers (1), CADD Manager's Office, CADD Stations (3), Drafting Stations (3), Purchasing Agent Purchasing Assistant, Pipe Superintendent, Sheet Metal Superintendent, Director of Marketing and Sales, Training Room, Conference Room (2), Copy and Fax Center, Supply Room, Bid Wrap-Up Room, Vice President of Estimating Office, Estimators (2), Administrative Assistant of Estimating, Library, Bid Take-off Room, Service Department, Warehouse Office, Lunchroom, Ladies Room, Men's Room, and Outdoor Courtyard

Second Level: Chief Financial Officer/Controller's Office, Accounting Department, Accounting Clerks (4), Human Resources, Exercise Room, and Computer Room

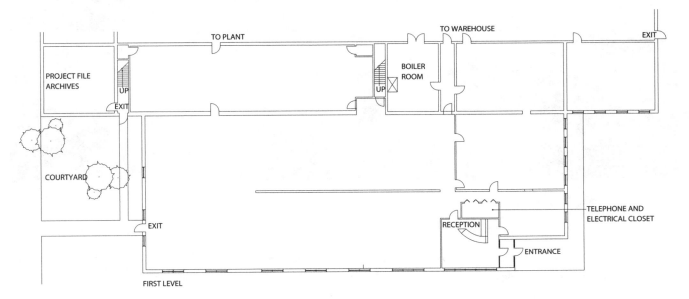

Figure 7–5 First-Level Plan

First Level Spatial Requirements

Entry Vestibule and Reception

Planning

- Square footage: 380. Adjacency: Bid Take-off Room
- The receptionist will monitor incoming and outgoing traffic of visitors and employees. Additional duties include typing, answering phone, taking messages, doing paperwork, receiving office deliveries, and sorting and processing mail. Visitors will be greeted, asked whom they wish to see, and wait in the reception area until someone arrives to receive them.
- Provide a seating area for visitors.
- Allow for adequate circulation space and a traffic path through the waiting area, to the office entry door, and to the Bid Take-off Room.

Design

- Give the visitors a good indication that they are working with a successful company.
- Materials must be of commercial grade, hold up under heavy usage, and meet codes. This applies to all areas unless otherwise noted.
- Woven or tightly tufted carpet with short level loop that meets ADA requirements will be used (maximum pile height of half inch). This applies to all areas unless otherwise noted.
- Reuse existing reception counter.
- Typical window dimension heights for the building are: floor to sill: 25″; sill to top of window: 75″.

Figure 7–6 Typical window elevation

Figure 7–7 Reception desk exterior. Color photo on CD.

Furniture

- Existing under counter file, box and pencil drawers must be removed; shelf remains.
- A retractable keyboard tray and pencil drawer to attach to existing transaction counter work surface
- Four under counter mobile pedestals with a file and two box drawers with locks
- Storage for stationery, forms, and multiple-size envelopes in two of the mobile pedestals
- Trays, dividers, and hanging folder bars for all file and box drawers
- Message slots for forty employees
- A task chair with arms and casters for carpet
- Task light at the desk
- Waste receptacle, paper trays, message holder, day calendar holder, artwork, and plants
- Seating arrangement for at least four people

Figure 7–8 Reception desk interior, left

Figure 7–9 Reception desk interior, center

Figure 7–10 Reception desk interior, right

- Side tables with lamps
- A magazine and bookrack, literature holders, and umbrella stand
- Coat storage
- Window covering

Equipment

- Typewriter
- Fax machine
- Computer and printer
- Telephone with an intercom and headset
- Emergency buzzer
- A security code keypad lock on the door between Reception and each entry door into the general office

INTERIOR CORRIDORS

Planning

- Develop circulation paths to move through the space.

Design

- Select materials and colors to address the heavy black soiling from warehouse and plant facility foot traffic.
- Provide general illumination, accent light for artwork and plants, illuminated exit signs, and emergency lighting.

Furniture

- Dispensing unit and trash receptacle for paper boot covers at all plant and warehouse entry doors
- Walk-off mat carpets at corridors leading directly to plant and warehouse doors; carpet to be contracted with a cleaning and rental service under a lease of quarterly replacements
- Artwork, plants, and any furnishings appropriate for the space

Equipment

- A security code keypad lock on the door into Reception, and plant and warehouse entry doors
- Fire extinguishers
- Water fountain

PRESIDENT

Planning

- Square footage: 470. Adjacency: Executive administrator and main conference room. Window location
- Plan an executive office suite with a private bath and enclosed wet bar area.
- Danna spends a lot of time with the vice presidents and clients and is involved in reviewing paperwork and plan files.
- Plan a conference area and soft seating group.

Design

- The President's office must be designed to make a positive impression on visitors and reflect a feeling that they are dealing with a successful company and individual. Danna has requested traditional casework.

- Private bath and wet bar area must have a hard surface floor covering.
- Provide window treatments.

Furniture

- A desk with legal file storage, pencil drawer, and box drawers, minimum size of 36″ × 72″
- A return or bridge surface that allows for maximum surface area
- A kneehole credenza with two lateral file drawers and storage capabilities
- Full wall storage units with filing capabilities; television cabinet; and wardrobe storage for coat, construction hat, and boots
- Executive, swivel/tilt chair with casters and leather upholstery
- Two guest chairs
- Trash receptacle, desk accessories, and task lighting
- Sofa and related furnishings
- Lamps
- Window covering, artwork, and plant
- Any furnishings that you find appropriate for the space

Equipment

- Computer
- Telephone
- Calculator/adding machine
- Wet bar: lavatory and under counter refrigerator
- Private bath: shower water closet, and lavatory

EXECUTIVE ADMINISTRATOR

Planning

- Square footage: 150. Adjacency: President's office
- Location requires easy access to the president and additional file space to hold records for the president.
- Communicates frequently with the president, all vice presidents, and clerical staff.
- Works approximately 8 to 10 hours a day and at least 2 of those hours are spent on the telephone. Majority of duties are performed on the computer, and about 1½ hours are split between fax, typewriter, and copy machine. Responsibilities include management of the clerical staff and typing and filing duties.
- Most documents handled are letter and legal size. The executive administrator is responsible for storage of all master copies of company manuals and preparation of insurance documents and administrative and confidential documents.
- If space allows provide a seating area for those waiting to see the president.

Design

- Furniture should be modular.

Furniture

- Desk-height work surfaces, including a retractable keyboard tray and pencil drawer
- Four 36″ wide lateral file drawers with locks
- Four legal size file drawers
- One pedestal with a file and two box drawers with locks
- Storage for stationery, forms, and multiple-size envelopes within a pedestal
- Trays, dividers, and hanging folder bars for all file drawers
- Overhead storage of at least 72 linear inches

- 20 linear feet of shelving for book and manual storage
- A task chair (without arms) with casters for carpet
- Tack surfaces required (approximately 6 square feet)
- Task light at the desk
- Guest seating for one
- Waste receptacle, paper trays, message holder, day calendar holder, artwork, and plants

Equipment
- Typewriter
- A computer with a printer
- Telephone with an intercom

MAIN CONFERENCE ROOM

Planning
- Square footage: 280. Adjacency: President's office
- The room will be used for client presentations, vendor presentations, executive staff meetings, and board meetings. Plan a conference table with the ability to seat twelve, including proper circulation spaces around the table.
- One wall will be used for a multifunction presentation cabinet and projection screen. Allow for additional circulation space in this area.
- Plan a beverage service area with storage for beverage serving items and conference room meeting items (paper, pens, pencil cups, and so on). This can be contained in a credenza or wall storage unit.

Design
- A well-appointed appearance complete with every detail is required.
- Design a multiple lighting system that includes dimming capabilities with multiple switch locations.

Furniture
- A conference table with electrical receptacles and connection capabilities for slide projector and two laptop computers
- Large comfortable swivel/tilt leather chairs with carpet casters for conference seating
- Wall presentation cabinet, including dry marker board, tackboard, and paper pads
- Floor easels and wall easel trim on the perimeter walls
- Projection screen with a minimum width of 8'0"
- Credenza storage and serving furniture components, including legal file drawers, box drawers, storage, and shelving
- Complete conference accessories, including individual table pads, portfolios, pencil cups, tablet holders, coasters, serving trays, water pitchers, waste cans, and so on
- Waste receptacles, artwork, and plants
- Window treatments
- Any furnishings that you find appropriate for the space

Equipment
- Computer and laptop LAN connections
- Telephone
- Television, slide projector, VCR, and DVD equipment
- Video equipment

VICE PRESIDENT OF FACILITY MANAGEMENT

Planning

- Square footage: 280. Adjacency: President's office; administrative assistant of Facility Management. Window location
- Private office

Design

- Furniture should be modular.
- Provide window covering.

Furniture

- Desk-height work surfaces (minimum 37″ × 72″), including a retractable keyboard tray and pencil drawer
- Conference table or work surface (minimum 8 linear feet)
- Filing units, including two letter file drawers and four legal file drawers, 12 linear feet of lateral file storage
- 12 linear feet of shelving
- Storage for construction hat and boots
- Executive chair
- Guest seating for four
- Large dry erase board and a 4′ × 4′ tack surface
- Storage for flat plans and rolled plans
- Literature storage bins for sorting project forms
- Waste receptacle, coat hooks, artwork, and plants

Equipment

- Calculator/adding machine
- Telephone
- Computer

ADMINISTRATIVE ASSISTANT OF FACILITY MANAGEMENT

Planning

- Square footage: 110. Adjacency: Vice president of Facility Management
- Communicates frequently with the vice president of Facility Management and clerical staff.
- Works approximately 8 hours a day and at least 2 of those hours are spent on the telephone. Majority of duties are performed on the computer, and about 2 hours are split between fax, typewriter, and copy machine. Responsibilities include typing and filing duties for the vice president of Facility Management. The administrative assistant communicates often with the superintendents and the Service Department.
- Most documents handled are letter and legal size. Responsible for preparation of service and facilities-related documents and forms.
- If space allows, provide guest seating.

Design

- Furniture should be modular.

Furniture

- Desk-height work surfaces, including a retractable keyboard tray and pencil drawer
- Four 36″ wide lateral file drawers with locks
- Four legal size file drawers

- One pedestal with a file and two box drawers with locks
- Storage for computer disks and CDs, stationery, forms, and multiple-size envelopes
- Trays, dividers, and hanging folder bars for all file drawers
- Overhead storage of at least 72 linear inches
- 4 linear feet of shelving for book and manual storage
- A task chair (without arms) with casters for carpet
- Tack surfaces required
- Task light at the desk
- Guest seating for one
- Waste receptacle, paper trays, message holder, day calendar holder, artwork, and plants

Equipment

- Shared typewriter (will be shared with other administrative assistants)
- A computer with a printer
- Telephone with an intercom

VICE PRESIDENT OF CONSTRUCTION

Planning

- Square footage: 270. Adjacency: Administrative assistant of Construction, project managers, and the president
- Responsible for project managers, purchasing, piping superintendent, and sheet metal superintendent.
- Plan for a private office with window location.

Design

- Furniture should be modular.
- Provide window covering.

Furniture

- Desk-height work surfaces (minimum 36" × 72"), including a retractable keyboard tray and pencil drawer
- Work surfaces for review of plans (minimum 10 linear feet) with shelving above; at least one portion of which must be an adjustable drafting surface with telephone shelf above
- Conference table or work surface (minimum 8 linear feet) for meetings and plans
- Filing units, including two letter file drawers and four legal file drawers, 12 linear feet of lateral file storage
- 26 linear feet of shelving
- Storage for construction hat and boots
- Executive chair
- Drafting stool
- Guest seating for four to be also used for conference table
- Large dry erase board and a 4′ × 4′ tack surface
- Storage for flat plans and rolled plans
- Literature storage bins for sorting project forms
- Waste receptacle, coat hooks, artwork, and plants

Equipment

- Calculator/adding machine
- Telephone
- Computer

ADMINISTRATIVE ASSISTANT OF CONSTRUCTION

Planning

- Square footage: 80. Adjacency: vice president of Construction and project managers
- Communicates frequently with the president, all vice presidents, and clerical staff.
- Works approximately 8 hours a day and at least 2 of those hours are spent on the telephone. Majority of duties are performed on the computer and about 2 hours are split between fax, typewriter, and 1 to 4 hours at the copy machine. Responsibilities include typing and filing duties for the vice president of Construction and the project managers. Communicates often with the service manager and occasionally with the superintendents.
- Most documents handled are letter and legal size. Responsible for preparation of project manuals and construction-related documents and forms. There is more shelving required to maintain manuals.
- If space allows, provide guest seating.

Design

- Furniture should be modular.

Furniture

- Desk-height work surfaces, including a retractable keyboard tray and pencil drawer
- Four 36″ wide lateral file drawers with locks
- Four legal size file drawers
- One pedestal with a file and two box drawers with locks
- Storage for computer disks and CDs, stationery, forms, and multiple-size envelopes
- Trays, dividers, and hanging folder bars for all file drawers
- Overhead storage of at least 72 linear inches
- 4 linear feet of shelving for book and manual storage
- A task chair (without arms) with casters for carpet
- Mobile typing stand
- Tack surfaces required
- Task light at the desk
- Guest seating for one
- Waste receptacle, paper trays, message holder, day calendar holder, artwork, and plants

Equipment

- Shared typewriter (will be shared with other administrative assistants)
- A computer with a printer
- Telephone with an intercom

PROJECT MANAGERS (FOUR REQUIRED)

Planning

- Square footage: 270 (for each PM). Adjacency: Administrative assistant to project managers and assistant project managers
- Because of the hectic pace of the construction industry, organization is the key to successful project management.
- The project manager (PM) writes the overall management "Plan" for his project, using a computerized system and tracking program. Project labor cost controls, and scheduling, combine with every trade in orchestrating the "Plan."
- PMs use a computerized system of tracking purchase orders, shop drawings, labor and material cost controls, and coordinating and scheduling processes. They also use the Estimating Department and its system to disseminate the piping and sheet metal material and labor.

- They work closely with the other trades to make the entire project run smoothly and minimize labor costs.
- Duties also include reviewing and processing submittal information, equipment and materials costs, and compiling and coordinating Critical Path Method (CPM) and Bar Chart schedules.
- Weekly meetings provide upper management a communication link with the field and project status reporting that keeps the project running smoothly. There are team meetings with other PMs, job superintendents, and field foremen to review project plan, drawings, and specifications.
- They will communicate very frequently with the Drafting Department for project coordination and cost control, and for the same reasons frequently communicate with the Estimating Department. There is also regular communication with the Accounting Department for cost control issues.
- They work an average of 10 to 11 hours per day and 3 hours of that time is spent on the telephone. They review and study construction plans, on three projects at one time. Blueprints (36″ × 48″), schedules, specifications, submittal information will be stored and handled in large quantities.
- Perks for PMs include a private office at a window location.

Design
- Furniture should be modular.

Furniture
- Desk-height work surfaces, including a retractable keyboard tray and pencil drawer
- Work surfaces for reviewing plans with shelving above, at least one of which must be an adjustable drafting surface
- Conference table or work surface
- Filing units, including two letter file drawers and four legal file drawers, 12 linear feet of lateral file storage
- 20 linear feet of shelving
- Storage for construction hat and boots
- Executive chair
- Guest seating for two
- Large dry erase board and a 4′ × 4′ tack surface
- Storage for flat plans and rolled plans
- Literature storage bins for sorting project forms
- Waste receptacle, coat hooks, artwork, and plants

Equipment
- Calculator/adding machine
- Telephone
- Computer

ASSISTANT PROJECT MANAGERS (TWO REQUIRED)

Planning
- Square footage: 175 (for each). Adjacency: Project managers and administrative assistant to project managers
- Assist project managers in reviewing and studying construction plans, and reviewing and processing submittal information.
- Plan for privacy and acoustical features to keep distractions to a minimum.

- Blueprints (36″ × 48″), schedules, specifications, and submittal information will be stored and handled in large quantities and often.

Design

- Furniture should be modular and provide privacy at each station.

Furniture

- Desk-height work surfaces, including a retractable keyboard tray and pencil drawer
- Work surfaces for reviewing plans with shelving above
- Filing units, including two letter file drawers and four legal file drawers
- 20 linear feet of shelving
- Storage for construction hat and boots
- Management level chair
- Guest seating for one
- Storage for flat plans and rolled plans
- Literature storage bins for sorting project forms
- Waste receptacle, coat hooks, artwork, and plants

Equipment

- Calculator/adding machine
- Telephone
- Computer

ADMINISTRATIVE ASSISTANT TO PROJECT MANAGERS (ONE REQUIRED)

Planning

- Square footage: 150. Adjacency: Project managers and administrative assistant of the vice president of Construction
- Provide a central area for assembly of project manuals.
- Communicates frequently with all project managers and assistant project managers.
- Works approximately 8 hours a day and at least 2 of those hours are spent on the telephone. Majority of duties are performed on the computer, and about 2 hours are split between fax and typewriter, and 1 to 4 hours at the copy machine. Responsibilities include typing and filing duties for the project managers.
- Most documents handled are plan, letter, and legal size. Responsible for preparation of project manuals and construction-related documents and forms. There is more shelving required to maintain manuals.

Design

- Furniture should be modular.

Furniture

- Desk-height work surfaces, including a retractable keyboard tray and pencil drawer
- Four 36″ wide lateral file drawers with locks
- Four legal size file drawers
- One pedestal with a file and two box drawers with locks
- Storage for computer disks and CDs, stationery, forms, and multiple-size envelopes
- Trays, dividers, and hanging folder bars for all file drawers
- Overhead storage of at least 72 linear inches
- 20 linear feet of shelving for book and manual storage
- Work surfaces for assembly of manuals and related equipment, including punch hole machines, binding machine, paper cutter, and blueprint mailing bags

- A task chair (without arms) with casters for carpet
- Two task chairs for manual assembly area
- Mobile typing stand
- Tack surfaces required
- Task light at the desk
- Waste receptacle, paper trays, message holder, day calendar holder, artwork, and plants

Equipment
- Calculator/adding machine
- Telephone
- Shared typewriter (will be shared with other administrative assistants)
- Computer and printer
- Central fax and copy machine area for PMs and Purchasing

PURCHASING AGENT

Planning
- Square footage: 150. Adjacency: Project managers
- The Purchasing Agent processes purchase orders for equipment and materials for projects.

Design
- Furniture should be modular and provide privacy at each station.

Furniture
- Desk-height work surfaces, including a retractable keyboard tray and pencil drawer
- Two 36″ wide, four-drawer lateral files with locks
- One pedestal with a file and box drawer with locks
- Overhead storage of at least 72 linear inches
- Minimum of 20 linear feet of catalog storage
- Shelf for literature storage bins for sorting project POs (provide at least fourteen bins)
- A management level chair with casters for carpet
- Guest seating for one
- Tack surfaces required
- Task light at the desk
- Artwork

Equipment
- Calculator/adding machine
- Telephone
- Computer and printer

PURCHASING ASSISTANT

Planning
- Square footage: 80. Adjacency: Purchasing Agent

Design
- Furniture should be modular.

Furniture
- Desk-height work surfaces, including a retractable keyboard tray and pencil drawer
- Two 36″ wide lateral file drawers with locks
- One mobile pedestal with a file and box drawer with locks
- Overhead storage of at least 72 linear inches

- A task chair with casters for carpet
- Tack surfaces required
- Task light at the desk
- Artwork

Equipment
- Calculator/adding machine
- Telephone
- Computer and printer

PIPE SUPERINTENDENT

Planning
- Square footage 125. Adjacency: Project managers
- Blueprints (36″ × 48″), schedules, specifications, and submittal information will be stored and handled in large quantities and often.

Furniture
- Desk-height work surfaces, including a retractable keyboard tray and pencil drawer
- Work surfaces for reviewing plans with shelving above, at least one portion of work surface must be an adjustable drafting surface
- Filing units, including two letter file drawers and two four-drawer lateral legal file units
- 20 linear feet of shelving
- Storage for construction hat and boots
- Management level chair
- Guest seating for one
- Large magnetic, dry erase board and a 4′ × 4′ tack surface
- Storage for flat plans and rolled plans
- Literature storage bins for sorting project forms
- Waste receptacle, coat hooks, artwork, and plants

Equipment
- Telephone
- Computer

SHEET METAL SUPERINTENDENT

Planning
- Same as for the pipe superintendent

Furniture
- Same as for the pipe superintendent

Equipment
- Same as for the pipe superintendent

CADD MANAGER

Planning
- Square footage: 155. Adjacency: CADD and Drafting Stations and project managers
- Responsible for supervision of CADD and Drafting Department and is the coordinator between project managers and all drafting employees.
- Communicates frequently with the project managers and occasionally with the plant supervisor. Communicates by telephone with the field foremen, architects, and engineers. The CADD manager occasionally meets with vendors and works approximately 8 hours a day.

- Works with plan documents 30″ × 42″ and 36″ × 48″.
- Provide the option to dim lighting to a low level during work on the computer.

Design

- Design a private office with modular furniture.

Furniture

- Desk-height work surfaces, including a retractable keyboard tray and pencil drawer; adequate surface area for computer and plans; reference surface a minimum of 36″ × 72″
- Plan table
- Four 36″ wide lateral file drawers with locks
- One pedestal with a file and two box drawers with locks
- Storage for computer disks and CDs, stationery, forms, and multiple-size envelopes
- Trays, dividers, and hanging folder bars for all file drawers
- Overhead storage of at least 72 linear inches
- 20 linear feet of shelving for book and manual storage
- A task chair with casters for carpet
- Tack surfaces required on all wall surfaces
- Task light at the desk
- Guest seating for two
- Waste receptacle, paper trays, message holder, day calendar holder, artwork, and plants

Equipment

- Computer and printer
- Wall-mounted telephone

CADD Stations (Three Required)

Planning

- Square footage: 115 (for each). Adjacency: CADD manager
- Communicates frequently with the CAD manager and project managers.
- Works approximately 8 hours a day.
- Works with plan documents 30″ × 42″ and 36″ × 48″.
- Prefers to have option to dim lighting to a low level while working on CADD computer program.

Design

- Furniture should be modular and provide privacy at each station.

Furniture

- Desk-height work surfaces, including a retractable keyboard tray and pencil drawer; adequate surface area for computer and plans; reference surface minimum of 36″ × 72″
- One pedestal with a file and two box drawers with locks
- Storage for computer disks and CDs
- Trays, dividers, and hanging folder bars for all file drawers
- Overhead storage of at least 72 linear inches
- 8 linear feet of shelving for book and manual storage
- A task chair with casters for carpet
- Plan rack for ten active projects
- Tack surfaces for all wall surfaces
- Task light at the desk
- Guest seating for one
- Waste receptacle, paper trays, message holder, day calendar holder, artwork, and plants

Equipment
- Computer
- Wall-mounted telephone

DRAFTING STATIONS (THREE REQUIRED)

Planning
- Square footage: 80 (for each). Adjacency: CADD manager
- Communicates frequently with the CADD manager, project managers, and occasionally with the plant supervisors.
- Works with plan documents 30″ × 42″ and 36″ × 48″.

Design
- Furniture should be modular and provide privacy at each station.

Furniture
- Traditional drafting table with storage cabinet
- Work surface with files below
- Reference surface minimum of 36″ × 72″
- One pedestal with a file and two box drawers with locks
- Storage for stationery, forms, and multiple-size envelopes
- Trays, dividers, and hanging folder bars for all file drawers
- Overhead storage of at least 72 linear inches
- A drafting chair with casters for carpet
- Plan racks
- Tack surfaces for all wall surfaces
- Task light at the desk
- Guest seating for one
- Waste receptacle, paper trays, message holder, day calendar holder, artwork, and plants

Equipment
- Electric pencil sharpener and eraser
- Wall-mounted telephone

PRINT ROOM

Planning
- Square footage: 300. Adjacency: CADD and Drafting Stations
- Plan an area to print and plot drawings.
- Provide storage for paper and printing and plotting supplies.

Design
- A hard surface floor covering is required.

Furniture
- Plan tables
- Five flat plan files
- Drafting supply storage cabinet
- 60 linear inches of shelving
- Storage for rolled drawings
- Tack surface
- Wall clock
- Any furnishings that you find appropriate for the space

Equipment

- Wall-mounted telephone
- Fire extinguisher, smoke detector, and first aid kit
- Plotter
- Blueprint machine and copy machine for oversized plan documents

DIRECTOR OF MARKETING AND SALES

Planning

- Square footage: 130. Adjacency: President
- Responsible for company sales and marketing, and is the company's representative for bid delivery and bid opening meetings.
- Communicates constantly with the president, very frequently with the estimating department for bidding and sales, and frequently with the executive administrator for clerical support.
- Works 10 to 11 hours a day and 50% of that time is spent on the telephone. Uses the fax and copy machines frequently, and spends a significant amount of time using the computer.
- Most documents handled are letter and legal size. Filing for 60″ letter and 36″ legal documents is required.
- Plan for a private office.

Design

- Prefers wood finish on furnishings.

Furniture

- Desk (minimum 72″ × 36″); two-drawer file pedestal and box drawer pedestal
- Credenza with two-drawer lateral file and storage
- Reference table
- 20 linear feet of book storage
- Executive chair
- Guest seating
- Storage for computer disks and CDs, stationery, forms, and envelopes
- Trays, dividers, and hanging folder bars for all file drawers
- Task lighting
- Waste receptacle and artwork
- Any furnishings that you find appropriate for the space

Equipment

- Telephone
- Computer and printer

TRAINING ROOM

Planning

- Square footage: 840
- Room is used for meetings, training, and presentations for employee groups of up to 50 persons.
- Plan a presentation area with a speaker's table and podium. Presentations will use audio and visual materials and equipment.
- Include classroom-style seating with narrow tables and stacking chairs.
- Provide equipment storage.

Design

- Design perimeter wall tack surfaces with easel ledge.

Furniture

- Seating for fifty people; stacking chairs requested
- Narrow folding tables (lengths of 10, 8, and 6 feet) to seat a minimum of thirty-six
- Head speaker's table to seat four
- Floor podium with electrical capabilities
- Audiovisual carts
- Pull-down presentation screen
- Closed cabinet dry marker board with tack surfaces and flip chart capability
- 20 linear feet of book shelving
- Telephone table
- Event board for entry wall
- Waste receptacles, movable coat storage, water cooler, and artwork
- Any furnishings that you find appropriate for the space

Equipment

- Wall-mounted telephone
- Overhead projector
- Slide projector
- Television
- VCR/DVD player
- Electric pencil sharpener
- Computer and laptop LAN connections
- Film projector

CONFERENCE ROOM (TWO REQUIRED)

Planning

- Square footage: 200 (for each). Adjacency: Reception
- Plan a conference table with the ability to seat six, including proper circulation spaces around the table.
- One wall will be the presentation wall for visual cabinet. Allow for additional circulation space in this area.

Furniture

- A conference table
- Large comfortable swivel/tilt chairs with carpet casters for conference seating
- Wall presentation cabinet, including dry marker board, tackboard, and paper pads
- Conference accessories, including pencil cups, coasters, and water pitchers
- Telephone table with drawer
- Waste receptacle, artwork, and plants

Equipment

- Telephone

COPY AND FAX CENTER

Planning

- Square footage: 110. Adjacency: Administrative assistants and Reception
- Plan a central copy area.
- Plan a central fax center.
- Plan a mail processing area.
- Plan an area with bins (forty) for in-house fax, memo, and mail distribution.

- Design storage for related work tools and supplies.

Furniture

- Any furnishings that you find appropriate for the space

Equipment

- Wall-mounted telephone
- Large copy machine
- Multifunction fax machine
- Postage machine

SUPPLY ROOM

Planning

- Square footage: 190
- Design a space for office supply storage.
- Provide storage for advertising brochures, and a secure room for company logo gifts.

Furniture

- 64 linear feet of shelving
- 8 linear feet of locking storage cabinets (20″ deep minimum × 84″ high)
- Any furnishings that you find appropriate for the space

BID WRAP-UP ROOM

Planning

- Square footage 290. Adjacency: Estimating Department
- This room is used during the final stages of the bidding process. The president, vice presidents, assigned estimator, and project manager are usually in attendance at these meetings.
- Plan a conference table to set eight people, including proper circulation spaces around the table.
- One wall will be the presentation wall. Allow for additional circulation space in this area.
- Plan an additional 24 linear feet of wall base cabinets at the perimeter. They should include storage for rolled plans and flat plans. Provide knee space at counter areas that are adjacent to the conference table.

Design

- There should be tackable wall covering on the presentation wall.

Furniture

- A conference table
- Large comfortable swivel/tilt chairs with carpet casters for conference seating
- Plan racks
- Waste receptacle, artwork, and plants

Equipment

- Three telephone locations

VICE PRESIDENT OF ESTIMATING DEPARTMENT

Planning

- Square footage: 180. Adjacency: Estimators and Bid Take-off Room
- Responsible for management of estimating department, reviewing and studying construction plans, including a total of three projects at one time for bidding purposes, and reviewing and processing bid submittal information.

- He or she communicates very frequently with engineers, architects, vendors, and the estimators for bid coordination.
- Works an average of 10 to 11 hours per day, and 5 hours of that time is spent on the telephone.
- Blueprints (36″ × 48″), schedules, specifications, and submittal information will be handled in large quantities and often.

Design

- Plan for a private office.

Furniture

- Executive desk with two box drawers and one file drawer
- Computer work surface return with retractable keyboard tray
- Credenza with two lateral file drawers
- Console table for display of awards
- Overhead shelf storage with tack surface below
- Executive chair
- Guest seating at the desk for two
- Storage for computer disks and CDs, stationery, forms, and envelopes
- Trays, dividers, and hanging folder bars for all file drawers
- Task lighting
- Waste receptacle, two paper trays, two vertical tray sorters, a diagonal tray sorter with six slots, a message holder, coat hooks, and artwork
- Any furnishings that you find appropriate for the space

Equipment

- Computer
- Printer
- Calculator
- Telephone with an intercom

ESTIMATORS (TWO REQUIRED)

Planning

- Square footage: 150 (for each). Adjacency: Vice president Estimating Department, and the Library
- Process involves reading plans and specifications, computerized information take-off, and estimating bids.
- An estimator works 8 hours a day and spends 1 hour on the telephone. Balance of time is spent at computer and calculator.
- Communicates frequently with the chief estimator.
- Documents handled are 36″ × 48″ plans and letter and legal size papers.
- Provide two surfaces for the layout of blueprint drawings.

Design

- Provide modular furniture with privacy panels.

Furniture

- Work surface of 60″ × 30″ with pencil drawer for telephone conversation and specification reviewing
- Take-off drafting surface (4′ × 5′) with 24″ extension for monitor
- Single pedestal with two box drawers and one file drawer with locks

- Two-drawer lateral file storage
- 8 linear feet of shelf storage
- Tack surface to display blueprints
- Task chair and drafting chair
- Guest seating for one
- Plan rack and storage for rolled plans
- Storage for computer disks and CDs, stationery, forms, and envelopes
- Trays, dividers, and hanging folder bars for all file drawers
- Task lighting
- Waste receptacle, paper tray, diagonal tray sorter with six slots, and artwork
- Any furnishings that you find appropriate for the space

Equipment

- Computer
- Take-off digitizer board
- Printer
- Calculator

ADMINISTRATIVE ASSISTANT OF ESTIMATING DEPARTMENT

Planning

- Square footage: 125. Adjacency: Vice president of Estimating Department and estimators
- Communicates frequently with the chief estimator, estimators, and vice president of Construction.
- Works approximately 8 hours a day and at least 1 hour is spent on the telephone. Majority of duties are performed on the computer and about 2 hours are split between the fax machine and typewriter, and 1 to 2 hours at the copy machine. Responsibilities include typing and filing duties.
- Most documents handled are plan size and letter and legal size. Responsible for preparation of bid submittal related documents and forms.
- Provide a central area for fax and copy machine.

Design

- Furniture should be modular.

Furniture

- Desk-height work surfaces, including a retractable keyboard tray and pencil drawer.
- Four 36″ wide lateral file drawers with locks
- Twelve vertical legal size file drawers
- Two pedestals with a file (legal) and two box drawers with locks
- Storage for computer disks and CDs, stationery, forms, and multiple-size envelopes
- Trays, dividers, and hanging folder bars for all file drawers
- Overhead storage of at least 72 linear inches
- A task chair with casters for carpet
- Two guest chairs
- Mobile typing stand
- Provide tack surfaces
- Task light at the desk
- Waste receptacle, paper trays, message holder, day calendar holder, artwork, and plants

Equipment

- Computer and printer

- Typewriter
- Telephone
- Fax machine

LIBRARY

Planning

- Square footage: 270. Adjacency: Estimating Department
- Plan for a resource library with catalog storage and reference manual storage.
- Plan space for a large copy machine and paper storage.

Furniture

- Provide 150 linear feet of shelving
- Library table
- Two chairs at the library table
- Waste receptacle, paper trays, pencil cups, and notepaper holders
- Any furnishings that you find appropriate for the space

Equipment

- Large copy machine

BID TAKE-OFF ROOM

Planning

- Square footage: 150. Adjacency: Reception and Estimating Department
- This room is used by vendors and field foremen for reading plans and specifications; information take-off for bidding by vendors; and material processing by foremen.
- Documents handled are 36″ × 48″ plans and letter and legal size paper.
- Provide at least three surfaces for the layout of blueprint drawings.

Furniture

- Desk-height work surfaces, including a retractable keyboard tray and pencil drawer
- Overhead open storage of at least 72 linear inches
- Two task chairs with casters for carpet
- Plan rack and storage for rolled plans
- Any furnishings that you find appropriate for the space

Equipment

- Computer
- Telephone

SERVICE DEPARTMENT MANAGER

Planning

- Square footage: 150. Adjacency: Warehouse and Plant
- The service manager supervises the start-up, service, and maintenance of all HVAC equipment installed. Commissioning and start-up technicians visit construction sites and discuss and review the system's progress with the installation foreman. After-market service, controls, and maintenance are also service programs performed.
- Works 8 hours a day and some of this time is spent tracking parts, material warranties, and so on.
- The service manager supervises eight to twelve field employees.
- Documents such as schedules and specifications will be stored and handled.

Design

- The office requires a hard surface floor covering.

Furniture

- Desk-height work surfaces, including a retractable keyboard tray and pencil drawer
- Plan review work surfaces with shelving above
- 144 linear inches of shelving
- Four letter file drawers and four legal file drawers
- Storage for flat plans and rolled plans
- Storage for construction hat and boots
- Management level chair
- Guest seating for one
- Large magnetic, dry erase board and a 4′ × 4′ tack surface for display of weekly scheduling and pending work
- Literature storage bins for sorting project forms
- Waste receptacle, coat hooks, artwork, and plants

Equipment

- Calculator/adding machine
- Telephone
- Computer

ASSISTANT TO THE SERVICE MANAGER

Planning

- Square footage: 120. Adjacency: Service manager
- Provide a central area for fax and copy machines.
- Communicates frequently with the service manager and the Warehouse and takes service calls.
- Works approximately 8 hours a day and at least 3 of those hours are spent on the telephone. Majority of duties are performed on the computer and about 1 hour is split between the fax machine, the typewriter, and the copy machine. Responsibilities include typing and filing duties.
- Most documents handled are letter and legal size.

Design

- Furniture should be modular.
- The office should have hard surface floor covering.

Furniture

- Desk-height work surfaces, including a retractable keyboard tray and pencil drawer
- Four 36″ wide lateral file drawers with locks
- Four legal size file drawers
- One pedestal with a file (legal) and two box drawers with locks
- Storage for computer disks and CDs, stationery, forms, and multiple-size envelopes
- Trays, dividers, and hanging folder bars for all file drawers
- Overhead storage of at least 72 linear inches
- A task chair with casters for carpet
- One guest chair
- Mobile typing stand
- Tack surfaces required
- Task light at the desk
- Waste receptacle, paper trays, message holder, day calendar holder, artwork, and plants

Equipment

- Computer and printer
- Typewriter
- Telephone
- Fax machine

WAREHOUSE OFFICE

Planning

- Square footage: 300. Adjacency: Loading dock, Warehouse, and Plant
- Staff duties include receiving equipment and materials, storing equipment and materials, and arranging for delivery of the same.
- Warehouse storage is a computerized tagging and tracking system.
- Three workstations are required.
- A meeting area is required.

Design

- A hard surface floor covering is required. Heavy black soiling occurs from plant foot traffic.
- Furniture should be modular.

Furniture

- Three workstations, including desk and computer surface, retractable keyboard tray, pencil drawer, pedestal with two box and one file drawer, and task chair
- Four letter size file drawers
- Table with four chairs
- Tack surface (4′ x 4′)
- 20 linear feet of catalog storage
- Storage for computer disks and CDs, stationery, forms, and multiple-size envelopes
- Trays, dividers, and hanging folder bars for all file drawers
- Waste receptacles

Equipment

- Two computers and one printer
- Fax machine and copy machine
- Three telephones

KITCHEN

Planning

- Square footage: 100.
- Design an area that opens to the Lunchroom.
- The Kitchen will be used to provide minimal food storage and preparation space for employees.
- The coffee and tea urns will be kept here.
- There will be storage and dispensing units for paper and plastic products, coffee, and related items.

Design

- Hard surface flooring is required in this area.

Equipment

- Cold storage for food and ice
- Dishwasher

- Ice maker
- Plumbing for a coffee maker
- Microwave
- Electrical receptacles at counter level

LUNCHROOM

Planning

- Square footage: 270. Adjacency: Kitchen
- Plan for a dining area, vending area, and seating arrangement.

Design

- Hard surface floor covering is required.

Furniture

- Tables and chairs to seat twelve
- Soft seating area for two to three people and related furnishings
- Two trash receptacles and tabletop accessories
- Artwork and plants
- Any furnishings that you find appropriate for the space

Equipment

- Soft drink machine: 72″ high × 41½″ wide × 31″ deep, with a black finish
- Candy machine: 72″ high × 40″ wide × 35½″ deep, with a black finish

LADIES ROOM

Planning

- Square footage: 116. Adjacency: Men's Room
- Handicap facilities are required.
- Provide lounge area.

Design

- Hard surface floor covering is required.

Furniture

- Mirrors, coat hooks, purse hooks, and shelves
- Soap dispensers, towel dispenser/waste receptacle, and sanitary napkin dispenser
- Seat cover, and toilet tissue dispenser with waste receptacles
- Reclining sofa and related furnishings
- Any furnishings that you find appropriate for the space

Equipment

- Partitions and grab bars
- Two water closets
- Two lavatories

MEN'S ROOM

Planning

- Square footage: 140. Adjacency: Ladies Room
- Handicap facilities are required.

Design

- Hard surface floor covering is required.

Furniture

- Mirrors, coat hooks, and shelves
- Soap dispensers, towel dispenser/waste receptacle
- Seat cover and, toilet tissue dispenser with waste receptacles
- Any furnishings that you find appropriate for the space

Equipment

- Partitions and grab bars
- One urinal
- One water closet
- Two lavatories

COURTYARD

Planning

- Arrange a dining area for six people.
- Provide seating in sunlight and in shade areas.

Design

- Materials must withstand heavy outdoor usage.

Furniture

- Table and chairs secured to the ground
- Three benches secured to the ground
- Waste and ash receptacles weighted for weather conditions
- Outdoor lighting for seating areas, walkways and entry doors

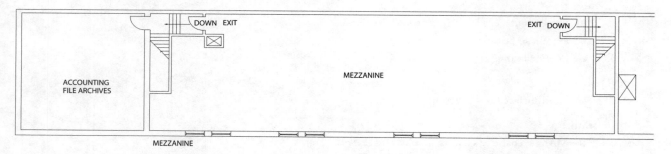

Figure 7–11 Mezzanine Floor Plan.

Second Level Spatial Requirements

CHIEF FINANCIAL OFFICER/CONTROLLER

Planning

- Square footage: 280. Adjacency: Accounting Department manager and Accounts Payable clerk
- Handles corporate level financial planning and monitors financial aspects of the corporation.
- Works 10 hours a day and 2 to 4 hours on weekends. Spends about 1 hour a day on the telephone.
- Communicates very frequently with the Accounting Department manager and the Accounts Payable clerk.
- Requires sufficient space to review large ledger books, computer reports, and financial worksheets for accounting. Most documents are 11″ × 17″ and 8½″ × 11″..
- All files must be secured for confidential corporate records.

Design

- Plan for a private office.

Furniture

- Desk-height work surfaces (72″ × 36″), including a retractable keyboard tray and pencil drawer
- Document review work surfaces 72″ × 48″
- Filing units, including two letter file drawers, two box drawers, and two four-drawer lateral legal file units, all with locks; four lateral drawers to hold electronic data processing (EDP) hanging files for computer reports
- 144 linear inches of shelving
- Storage for computer disks and CDs, stationery, forms, and multiple-size envelopes
- Trays, dividers, and hanging folder bars for all file drawers
- Executive level chair
- Guest seating for four at the conference table
- Large magnetic, dry erase board.
- Six paper tray holders
- Waste receptacle, coat hooks, artwork, and plants

Equipment

- Computer with printer
- Adding machine
- Telephone

ACCOUNTING DEPARTMENT MANAGER

Planning

- Square footage: 125. Adjacency: Controller
- Responsible for department scheduling and daily work performance of staff.
- Works 8 to 9 hours a day and communicates very frequently with the controller, and regularly with the accounting clerks.
- Handles letter, legal, and 11″ × 17″ computer documents.

Design

- Furniture should be modular and provide visual and acoustical privacy.

Furniture

- Desk-height work surfaces, including two retractable keyboard trays and a pencil drawer
- Two document review work surfaces
- Filing units, including two lateral letter file drawers, two lateral legal file drawers, and one pedestal with two box drawers and one legal file drawer, all with locks; legal files storage with front to back orientation; rolling file bin with hanging file capabilities for data binders
- 8 linear feet of shelving
- Storage for computer disks and CDs, stationery, forms, and multiple-size envelopes
- Trays, dividers, and hanging folder bars for all file drawers
- Management level chair
- Eight paper tray holders and two mini shelf units with dividers
- Waste receptacle, coat hooks, artwork, and plants
- Mobile typewriter station

Equipment

- Adding machine
- Two computer terminals and one printer
- Telephone

ACCOUNTING CLERKS (THREE REQUIRED)

Planning

- Square footage: 125. Adjacency: Accounting Department manager
- The accounting clerk processes accounts payable and payroll, and distributes reports.
- Works 7 to 8 hours a day and communicates with the controller and Accounting Department manager and accounting clerks.
- Handles letter, legal, and 11" × 15" computer documents.
- All processes are computerized. Fax and copy machines are used frequently as well.
- A central filing area and shared typing station are required.

Design

- Furniture should be modular and provide privacy at each workstation.

Furniture

- Desk-height work surfaces, including retractable keyboard trays and two pencil drawers; two retractable keyboard trays at the payroll station; and document review work surfaces
- Filing units, including two lateral letter file drawers, one pedestal with two box drawers and one legal file drawer, all with locks; legal files in front to back orientation; rolling file bin with hanging file capabilities for data binders; ten rolling file bins at accounts payable station
- 8 linear feet of shelving
- Storage for computer disks and CDs, stationery, forms, and multiple-size envelopes
- Trays, dividers, and hanging folder bars for all file drawers
- Task chair with arms and carpet casters
- Eight paper tray holders, diagonal tray holder with three slots, two vertical bin organizers, and two mini shelf units with dividers
- Waste receptacle, coat hooks, artwork, and plants
- Mobile typewriter station shared and stationed at the Accounting Department manager's station

Equipment

- Computers with printer
- Laser printer at Accounts Payable station
- Adding machine
- Telephone

ACCOUNTING FILE AREA AND ACCOUNTING LIBRARY

Planning

- Square footage:420. Adjacency: Accounting clerks
- Space is required to review large ledger books and sheets for accounting.
- A central filing area is required for accounting records.
- A central copy and fax center with wall telephone nearby is required.
- Financial books, tax manuals, and operational manuals will be kept in the Library.

Design

- Open area

Furniture

- 12 linear feet of EDP file storage with flipper doors and a T-bar suspension in the interior of the file
- 24 linear feet of legal file drawers
- 100 linear feet of letter file drawers
- Library table with seating for two

- 40 linear feet of shelving for books
- Coat storage
- Polar water dispenser
- Any furnishings that you find appropriate for the space

Equipment

- Fax and copy machines at the central location
- Telephone

HUMAN RESOURCES

Planning

- Square footage: 240
- She or he communicates with the president and other department heads regarding employment issues and frequently with the executive administrator for clerical support.
- Works 8 hours a day and 20% of that time is spent on the telephone. Uses fax and copy machines frequently and spends a significant amount of time using the computer.
- Most documents handled are letter and legal size. Filing for 60″ of letter and 36″ of legal documents.
- Private office

Furniture

- Desk (minimum 72″ × 36″), with a two-drawer file pedestal, and a box drawer pedestal
- Credenza with a two-drawer lateral file and storage
- Reference table
- 8 linear feet of book storage
- Executive chair
- Guest seating for two people
- Storage for computer disks and CDs, stationery, forms, and envelopes
- Trays, dividers, and hanging folder bars for all file drawers
- Task lighting
- Waste receptacle and artwork
- Any furnishings that you find appropriate for the space

Equipment

- Telephone
- Computer and printer

EXERCISE ROOM

Planning

- Square footage: 320. Adjacency: File archives.
- Design an exercise and workout area for office employees. Room is not supervised and will be used after work.

Furniture

- Storage cabinet
- Mirrors
- Ballet bars
- Seating
- Weight bench
- Three exercise mats
- Any furnishings that you find appropriate for the space

Equipment
- Television with VCR/DVD
- Pilates equipment
- Free weights
- An aerobic equipment machine such as a treadmill or stair climber

COMPUTER ROOM

Planning
- Square footage: 60. Adjacency: Accounting Department
- This is the location of the computer server and where computer software and manuals will be stored.
- Provide lock on door.

Furniture
- Cabinet for computer server
- Work surface for monitor and keyboard
- Task chair

Equipment
- Server, monitor, and keyboard

SUBMITTAL REQUIREMENTS

Phase I: Programming

Part A Due: _____   Part B Due: _____   Part C Due: _____

Part D Due: _____   Part E Due: _____   Part F Due: _____

A. Your instructor will organize the design team (space planner, merchandiser, designer, color and material designer) and negotiate a compensation method. Verify each team member's ability to meet the time schedule for the project.

B. Research various shapes, sizes, codes, arrangements, and clearances for the above-outlined spaces. Research fixture, furniture, and equipment for the same.

C. Hold two design team meetings, including a buzz session and dialogue group for the project. In the buzz session, discuss the topic of space planning, and propose alternative strategies. In the dialogue group session, share your knowledge from the research you have completed and arrive at a design approach for the project.

D. Analyze the program outlined above and develop a programming matrix keyed for adjacency locations. A sample blank matrix is provided in Chapter 31.

E. With the information gathered from the matrix, provide a minimum of three bubble diagrams for the spatial relationships of the areas required.

F. Provide a one-page typed concept statement of your proposed planning strategy and initial design concept approach.

Presentation formats for Parts A through F: White bond paper, 8½″ × 11″, with a border. Insert in the binder or folder. The matrix and bubble diagrams must be completed on the CADD program.

Evaluation Document: Programming Evaluation Form

Phase II: Conceptual Design   Part A Due: _____
Part B Due: _____   Part C Due: _____   Part D Due: _____

A. Provide preliminary partition and department location plans, showing all walls, ceilings, built-in fixtures, doors, and windows. Label departments and rooms and indicate the square footage required and the square footage planned. These plans can be freehand drawings or CADD drawings in 1/8″ scale. Drawings must have an appropriate border. Finished drawings must be a blueprint or white bond copy.

B. Develop the above partition plans, including a preliminary ceiling, furniture, fixture, and furnishings layout of each department and room. Label departments and rooms and indicate the square footage required and the square footage planned. This plan can be a freehand drawing or CADD drawing in 1/8″ scale. Drawings must have an appropriate border. Finished drawings must be a blueprint or white bond copy.

C. Develop the above plans into final floor plans, showing all items noted in the program. This plan must be completed using the CADD program. Develop ceiling and lighting studies into a final reflected ceiling plan, showing all ceiling conditions and heights on the CADD program.

D. Discuss the following mock preliminary plan check with your instructor before proceeding. Assign a plan check manager from your group who must arrange a meeting with your local authorities for a preliminary plan checking procedure as it applies to life safety and accessibility requirements for your project. As a group, have the above floor plans reviewed for code compliance. In the event your local authorities are not able to accommodate a school group request such as this one, ask a code official or architect to visit your class and share his or her knowledge in reviewing your plan.

Evaluation Document: Conceptual Design Evaluation Form

Phase III: Design Development
Part A Due: _____   Part B Due: _____

A. Develop a color, material, and a final furniture scheme. Samples and pictures can be presented in a loose format.

B. Develop perspective sketches and/or elevations of key design elements, fixtures, and furnishings.

Evaluation Document: Design Development Evaluation Form

Phase IV: Final Design Presentation   Date Due: _____

A. The final presentation will be an oral presentation to a panel jury. The project must be submitted on 20″ × 30″ illustration boards excluding the fixture plans and reflected ceiling plans. Your name, the date, the project name and number, the instructor's name, and the course title must be printed on the back of each board in the lower left-hand corner. Submit as many boards as required, including a colored elevation or rendering, and fixture, furniture, and material selections of the following (as assigned by the instructor):

1. _____
2. _____
3. _____
4. _____
5. _____
6. _____

B. The plans should be presented in 1/8″ scale. The elevations should be presented in 1/2″ scale.

Evaluation Document: Final Design Presentation Evaluation Form

Project Time Management Schedule: Schedule of Activities, Chapter 31.

The AutoCAD LT 2002 drawing file name is danna.dwg and can be found on the CD.

REFERENCES

See previous office projects for listings of books.

Trade Magazines

Contract Magazine: http://www.contractmagazine.com
Architectural Record: http://www.architecturalrecord.com
Metropolis Magazine: http://www.metropolismag.com

Product Manufacturers

Office Specialty: http://www.officespecialty.com
Haworth: http://www.haworth.com
Kimball International: http://www.kimball.com
Details, Ergonomic Office Accessories: http://www.details-worktools.com
Vecta: http://www.vecta.com
SitOnIt Office Seating: http://www.sitonit.net
Halcon: http://www.halconcorp.com
Groupe Lacasse: http://www.groupelacasse.com
Teknion: http://www.teknion.com
Tella: http://www.tellainc.com
Amuneal Manufacturing Corp.: http://www.amunealmetalforms.com

NOTES

RETAIL
FACILITIES

MALE LOCKER ROOMS

RECEIVING DOOR

SPACE #R-245

PATTERN
SMOKE GRAY

CHAPTER 8

Intuitive Wireless

OUTCOME SUMMARY

The project focuses on applying the design process to a service retailer. It also focuses on analysis and "techniques for ideation[1]," including role-playing, brainstorming, and synectics. Requirements are developing a schematic, or initial prototype design, including adjacency analysis to establish optimal circulation and use. These ideas will be communicated through drawings, sketches, material samples, and space plans.

client profile

In this scenario you will be developing a retail center prototype for Intuitive Wireless. "Self-expression" is the catchphrase for the company, because they encourage their customers to be creative with the wide-ranging possibilities of the wireless services they offer. Perceptive enough to know that each customer has individual needs, they concentrate on offering numerous options made obtainable by the sophisticated technological resources at hand. The products are Digital Wireless, Analog Wireless, and Wireless Data, which include Short Messaging, Internet, and Interactive Paging.

This project is a joint venture, and as the retail expert team member, your responsibility is to develop several planning and design concepts (rough space planning and concept sketches) for the stores. Louis Aguilar and Ken Smith are the architectural team members from MGA Architects who have asked you to join the design group.

Memorable, unique, elegant, and informative are words the client used to describe the design of the store. The spirit of the interior and the image projected from it must reveal the client's viewpoint and have an identifiable persona. It is especially important that it be a comfortable environment where one can wander and explore the unlimited combination of options. You must research and discover what atmosphere would be persuasive and appealing to the clientele. The client has done a considerable amount of situational research by visiting competitors and investigating point of purchase fixtures and other service retailers. Their feedback is that they do not want to look like the competitors. The Microsoft fixtures they viewed at a computer retail outlet left a positive impression in terms of fixture design as well as some banner concepts on both the interior and exterior of stores, which they would like to use for promotions.

The customer group is defined by "lifestyle," and the marketing approach addresses each individual's needs. As a destination store they wish to attract a higher-end customer with funds and the different occasions in which to splurge. It is the client's intent that the customer will interact with the products displayed. The user groups defined by lifestyles are weekend users, emergency users, and the business and executive user. But the overall focus of the operation and merchandise is geared toward the executive user group. The major promotions are seasonal and the client anticipates that the sales for promotions, such as graduation and Father's Day, will be purchase contracts, which are activated at a later date.

---

1. Kilmer, Rosemary & W. Otie (1992). Designing Interiors. Fort Worth, Texas: Harcourt Brace Jovanovich College Publishers. p 167

The service activities and their functions planned for Intuitive Wireless retail centers include:

Activation Process: Includes credit check, selection of rate plans, mobile phone activation, selection of NPA NXX and optional services.

Billing: Includes billing inquiry (view only), posting of credits and adjustments, taking deposits in cash or by check, changing rate plan, adding or canceling service, changing payment method and accepting checks, accepting cash and credit cards for bill payment.

Equipment: Includes sales, repair, or replacement of in-warranty and out-of-warranty equipment, exchange, troubleshooting, replacement of parts, loaner equipment, and demo equipment.

Accessories: Include product sales (batteries, chargers, Pre-Pay cards, portable hands-free kits, cases), exchange, and troubleshooting.

Coverage: Includes service coverage inquiries and display of coverage areas.

Enhanced Services: Include selling wireless data and fax, text messaging/enhanced text messaging services for wireless, voice mail services for wireless, insurance for wireless equipment, Smart Rescue for wireless subscribers, voice mail services for landline, ISDN lines, Internet (landline connection), and long distance for landline.

SIM: Includes the sale of SIM cards, loading data onto SIM cards (e.g., speed dial list), and troubleshooting.

Contact Tracking: Includes entering notes into database for tracking by customer account number.

Troubleshooting: Includes customer service in the areas of billing, equipment, SIM cards, enhanced services, and inbound roaming customers.

Rentals: Include rentals for equipment and subscriptions for visitors to use on Intuitive network: rental of equipment only for international GSM subscribers using their own SIM card; and rental of equipment and/or SIM for Intuitive subscribers to take abroad.

Pre-Pay Cards and Deposit Programs: Include selling pre-pay cards and taking deposits.

Intuitive Service: Includes selling Intuitive Internet and referring customers to appropriate Intuitive entities for service (ISDN, Answering Center, bill payment, Directory, etc.).

Tracking: Includes tracking sales for commission payments/compensation and leads management.

project details

General and Architectural Information

Location: To be determined by real estate division who is seeking stand-alone stores in all of California

Gross square footage: There will be two sizes for the stores. Develop two prototype plans based on approximately 1,300 and 1,500 square feet. The retail portion will remain the same in both stores and the back areas will be reapportioned by percentage as required.

Initial number of employees: Two to three. Projected number of employees: Four within a year

Hours of operation: Generally Monday–Friday: 10 a.m. to 9 p.m.; Saturday: 10 a.m. to 7 p.m.; Sunday: 11 a.m. to 6 p.m. Weekend hours could vary depending on location.

New construction: Bidding is required.

Consultants required: Architect, electrical engineer, and mechanical engineer

Security: Built in Sensormatic equipment is required at the entrance. Security cameras will be used. Lock devices are not required on fixtures, excluding the cash drawers.

Codes and Governing Regulations

Occupancy: Mercantile or Business

Because no specific site has been chosen, design the project to comply with the most current edition of the ICBO Uniform Building Code (UBC), the National Electrical Code (NEC), the

Uniform Fire Code (UFC), the California Building Code (CBC), and the Americans with Disabilities Act. Special seismic life safety issues apply to construction because California is in a risk level 4 seismic zone. The UBC contains information from the seismological and structural engineering communities to reduce the risk of damage from earthquakes. Use materials with the highest smoke and fire ratings.

An accessible cash wrap, a minimum of two exits from the space, and illuminated exit signs throughout are required.

Facilities Required

Store Front, Mobile Phone Displays, Info Center, User Sites, Product Fixtures, Training and Activation, Cash Wrap, Credit, and Service. Non-selling areas include Restroom, Stockroom, and Mechanical Service Equipment Room.

Spatial Requirements

STOREFRONT

- Establish an exclusive and highly identifiable design statement for the storefront, which relates to the interior concept.
- Most stores will be located in neighborhood strip centers. The customer will approach from a parking lot in that case. Other locations could include mall kiosks or a stand alone store in a shopping complex.
- Elements include entry door, show window, and an engaging graphic identification.
- Provide a show window with the capability of displaying products, graphics, and promotions. The show window must have visibility into the store.

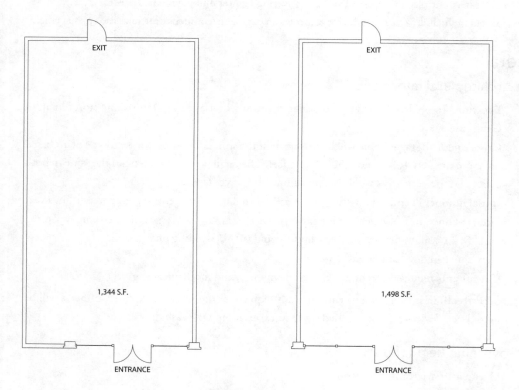

Figure 8–1 Suggested Floor Plans

MOBILE PHONE DISPLAYS

- Adjacency: User Sites
- Organize and present according to lifestyle, and provide one mobile phone display at the entrance.
- Mobile phone displays must include a product mobile phone(s) and literature for the product (two sizes). Typical brochure sizes are: 8½″ × 11″ triple fold and 8½″ × 11″ double fold. Include six or seven types of mobile phones made by four different manufacturers.

INFO CENTER

- Adjacency: Training and Activation
- The Info Center will explain the technology and applications for its use.
- It must be a highly active and strong area visually but, most importantly, be informative. The salespeople are very service oriented and will be using this area to inform and discuss services with the customer.
- Make it fun for the customer by providing interactive capabilities with three or four activated mobile phones and connections to the Internet.
- Incorporate original graphics and video capabilities. The client wants to use a large coverage map here that will be changed once a year. Plan for its location and size, and determine finish and construction materials. This element is to be dramatic and enticing and include eye-catching elements and innovative materials.

USER SITES

- Adjacency: Mobile Phone Displays
- Provide three or four User Sites that are organized according to lifestyle: weekend, business, and executive users. The flow should be from weekend to executive user.
- These are to be important meeting places where the customer and the sales staff come together to discuss a service or product. The sites will be used to gather facts, disseminate information to the customer, and give detailed explanations of the service, products, and plans. The Mobile Phone Displays will be used in this process and it is imperative that they are close to the User Sites.
- Important information to be presented and displayed will be the cover maps, rate plans, and equipment. Typical brochure sizes are: 8½″ × 11″ triple fold and 8½″ × 11″ double fold.
- Consider comfort for conversation and incorporate a work surface and chairs for discussions with the customer.

PRODUCT FIXTURES

- Design a fixture(s) combining shelving, bins and short peg face outs with optimum flexibility and mobility. Incorporate graphic capability to identify products that is changeable for seasonal promotions. Provide bins for product literature as part of the fixture function and design. Typical brochure sizes are: 8½″ × 11″ triple fold and 8½″ × 11″ double fold.
- There are six or seven types of mobile phones made by four different manufacturers, including numerous accessory items.
- The products include both flip and non-flip handsets. The accessories include batteries, vehicle power adapters, leather pouches and cases, watertight pouches, multichargers, travel chargers, rapid travel chargers, compact charging stands, rapid battery chargers, rapid cigarette lighter chargers, desktop chargers, executive kits, portable hands-free headsets, belt clips, silent call alert, car accessory kits, windshield mounts, antennas, starter kits, retractable antennas, and mobile holders. Other products include SIM cards, pre-pay cards, T-shirts, hats, and other trinkets.

- Dedicate an area/fixture for discontinued sale merchandise or excess supply.

TRAINING AND ACTIVATION

- Adjacency: Info Center and Cash Wrap
- Special services require training for use. The focus of this area is selling enhanced features to executive users, training for using special services, and mobile phone activation.
- A laptop computer and plug-ins for mobile phones are required for these areas.
- Provide seating because training and mobile phone activation require time and focus.

CASH WRAP

- Adjacency: Activation and Credit
- The area is used to write up sales and wrap the merchandise. Storage is required for bags and miscellaneous business items.
- Electrical equipment in this area will be two computerized cash terminals, hand scanner, calculator/adding machine, credit approval, payment machines, and two telephones.
- ADA requires that cash wraps be on an accessible route and double cash wraps are to be 36″ apart. A section of the customer side of the cash wrap must meet ADA guidelines.
- Develop a creative solution incorporating counter surfaces, storage and functional needs.

CREDIT

- Adjacency: Cash Wrap
- Sales contracts are completed in this area, and credit checks are performed before the on-site activation process. Additionally, it will function for bill inquiry, to post credits and adjustments, change rate plans, add or delete a service, and change payment methods.
- Electrical equipment in this area will be a computer, calculator/adding machine, credit card terminal, and telephone.
- A work surface of four linear feet, task chair, two chairs for customers, and file storage are required. A section of the customer side of the work surface must meet ADA guidelines.

SERVICE

- Locate at the back of the sales space.
- Minor repairs are performed in this area, and loan or rental mobile phones are provided here.
- Provide a work surface, counter stool, storage, and shelving.

RESTROOM

- The restroom is primarily for employees and secondary for customers and must meet ADA guidelines.
- Storage is required for employee belongings and cleaning materials.
- Provide one water closet and a hand-washing sink.

STOCKROOM

- This area is for back stock of products and supplies. Shelving is required as well as one locking storage cabinet. Provide a safe 24″ × 24″ × 24″.
- If space allows, provide a work surface, chair, and file storage for the store manager.

EQUIPMENT ROOM

- Allow space for electrical, telephone, and plumbing equipment.
- Provide a janitor sink and storage closet.

SUBMITTAL REQUIREMENTS

Phase I: Programming

Part A Due: _____   Part B Due: _____   Part C Due: _____

A. Analyze the program outlined above and develop a programming matrix keyed for adjacency locations. A sample blank matrix is provided in Chapter 31.

B. With the information gathered from the matrix, provide a minimum of three bubble diagrams for the spatial relationships of the areas required.

C. To generate ideas for the next phase, use the following techniques:

1. Role-playing. With the instructor as the leader do this exercise as a group session.
2. Brainstorming. With the instructor as the leader do this exercise as a group session.
3. Synectics.[1] Do this exercise as an informal interchange within a group.
4. Research the competition.

Presentation formats for Parts A through C: White bond paper, 8½″ × 11″, with a border. Insert in the binder or folder. The matrix and bubble diagrams are to be completed on the CADD program.

Evaluation Document: Programming Evaluation Form

Phase II: Conceptual Design

Part A Due: _____   Part B Due: _____   Part C Due: _____
Part D Due: _____   Part E Due: _____

A. Develop a preliminary partition and department location plan showing all walls, ceilings, built-in fixtures, doors, and windows. Label departments and rooms and indicate the square footage planned. These plans can be freehand drawings or CADD drawings in 1/4″ scale. Drawings must have an appropriate border. Finished drawings must be a blueprint or white bond copy.

B. Develop the storefront elevation and sections, showing all design and signage elements using any desirable door location, window elevation, or materials.

C. Develop the above plans into a final floor plan, showing all items noted in the program. This plan is to be completed using the CADD program. Develop ceiling and lighting studies into a reflected ceiling plan, showing all ceiling conditions and heights on the CADD program. Drawings must have an appropriate border. Finished drawings must be a blueprint or white bond copy.

D. Develop a color, material, and fixture scheme. Samples and pictures can be presented in a loose format.

E. Develop a perspective sketch of the interior of the store, showing all design elements and furnishings. Refine the storefront elevation and sections, showing all design and signage elements. Apply partial color to the perspective and elevations.

Evaluation Document: Conceptual Design Evaluation Form, Design Development Evaluation Form, and Final Design Presentation Evaluation Form

Project Time Management Schedule: See Schedule of Activities, Chapter 31.

The AutoCAD LT 2002 drawing file name is intuitive.dwg and can be found on the CD.

---

1. Synectics is a creative thinking technique to solve a problem through the use of metaphors and analogies. The primary goal is to develop ideas that will be innovative and unique.

REFERENCES

Special Resources

Synectics example:

Kilmer, Rosemary, and W. Otie. 1992. *Designing interiors.* Forth Worth, Tex.: Harcourt Brace Jovanovich College Publishers, p. 168.

Trade Magazines

VM&SD: http://www.visualstore.com

Stores Magazine: http://www.stores.org

Trade Organizations

Institute of Store Planners: http://www.ispo.org

National Association of Store Fixture Manufacturers: http://www.nasfm.org

National Retail Federation: http://www.nrf.com

International Council of Shopping Centers: http://www.icsc.org

Fixture Manufacturers

ALU: http://www.alu.com

RHC| Spacemaster Corporation: http://www.rhcspacemaster.com

Capitol Hardware: http://www.capitolhardwareinc.com

Goer Manufacturing Company, Inc.: http://www.goer.com

NOTES

CHAPTER 9

The Tango Club

OUTCOME SUMMARY

The project features four phases of the design process: identifying and analyzing the client's needs and goals, developing conceptual skills through schematic or initial design concepts, detailing and refining ideas from the schematic design phase, and drafting documents in preparation for the bidding and contracting of construction, fixtures, and furnishings. You are responsible for reviewing the landlord/tenant design criteria manual as it applies to the storefront and sign design. The tenant design criteria manual section focuses on exposure to retail tenant requirements and increasing knowledge of that procedural practice.

client profile

Vicky Marc, a Newport Beach based visual merchandiser and store owner, is well versed in retail trends and wants you to design a knockout boutique called The Tango Club. Her idea is for a boutique with a party atmosphere that emulates a New York City nightclub, and she promises Latin jazz, videos, and art ware fashions. Marc has a very hip design sensibility. The shop is to be located in Coral Isle Shopping Center. Taking her cues from designers such as Rei Kawakubo of Comme des Garcons, the merchandise will be forward thinking and the mood of the store the same. "Key items of furniture, slabs of stone, the sculptural treatment of display racks, and a characteristic use of lighting supply the distinctive Kawakubo signature."[1]

After visiting Kawakubo's Wooster Street, New York store, Marc's viewpoint for the shop is to link the visual expression of New York's SoHo with the energy and attitude of southern California. Key elements that she wants to incorporate are the architectural qualities of concrete and stone, large low tables, and slab shelving.

The plan needs to be logical, and the fixtures designed with modular flexibility. All perimeter (wall) and floor (loose) fixtures are to be a custom and creative design solution. The preferred standard spacing is 2'0" on center, and adequate mirrors for previewing garments must be placed strategically throughout the store. No two-way (also called T-stands), four-way, or round metal racks are to be used in the store.

For maximum flexibility, the use of track lighting is preferred. Lighting fixtures are to be energy efficient and meet California Title 24

Figure 9–1   Business Card and Logo

1. Sudjic 1990.

regulations, which restrict wattage to a total of approximately 3 watts per square foot. The maximum ceiling height is 12′0″ (minimum 8′0″).

The graphic identity for the shop and the logo as shown in Figure 9–1 will appear on the bags, tissue stickers, and promotional materials. The colors of the logo are black and white. The leopard skin is tan and dark warm gray and lends an exotic air to

Figure 9–2   Hat boxes

the theme. Marc has commissioned a French artist for the entertaining ingredient of adding a layer or two of lively visuals to the shop. The large colorful hatboxes with Tango letters and bold nude wall mural (6′ × 4′) he has envisioned will give the place a noticeably Marc energy. Are you in the mood yet? Meet you at Coral Isle.

Designed to look like a Mediterranean type of village, Coral Isle is located in Newport Beach, very close to the Pacific Ocean. The average income of residents of the area is over $120,000, with ample time to shop. It is a super regional shopping center that was built in 1967 and now boasts over 1,000,000 square feet of retail with anchor tenants, including Neiman Marcus, Bloomingdale's, Macy's, and Robinson's. When shopping at The Tango Club you will find yourself mixing with BCBG, Kenneth Cole, BeBe, L'Occitane, Betsey Johnson, Shabby Chic, Anthropologie, Nike Goddess, Apple Stores, and Z Galleries. Marc has traveled the globe to assemble an up-to-the-minute collection of clothing. The Tango Club will offer apparel by designers from Paris, London, Denmark, Hong Kong, Italy, and California, and the target customer will be like-minded.

project details

General and Architectural Information

Location: The Tango Club, 24 Coral Isle, Tenant Space #313

Gross square footage: 2081. Net usable square footage: 2016

Initial number of employees: Two. Projected number of employees: Four within 1 year

Hours of operation: Monday–Friday: 10 a.m. to 9 p.m.; Saturday: 10 a.m. to 7 p.m.; Sunday: 11 a.m. to 6 p.m.

New construction: Bidding is required. Mall facility plans are available. Site information gathering is required.

Consultants required: Architect, electrical engineer, and mechanical engineer

Security: Built-in Sensormatic equipment is not required at the entrance. Security cameras will be used, and the mall provides 24-hour security. Lock devices are not required on fixtures, excluding the showcases and cash drawers.

Codes and Governing Regulations

Occupancy: Mercantile

Building type: Type II. Fire Resistive

Design the project to comply with the most current edition of the ICBO Uniform Building Code (UBC), the National Electrical Code (NEC), the Uniform Fire Code (UFC), California

Figure 9–3 Lease Plan

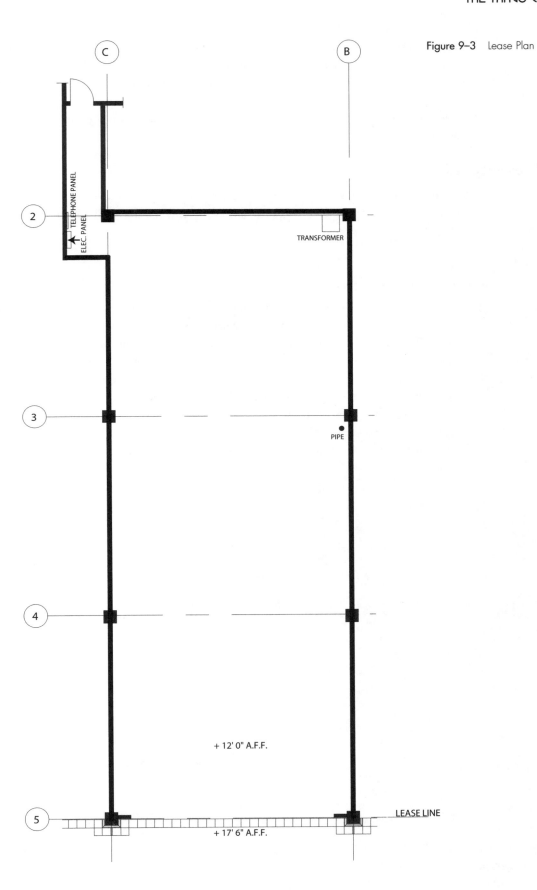

Building Code (CBC), and the Americans with Disabilities Act. Special seismic life safety issues apply to construction because the structure is in a risk level 4 seismic zone. The UBC contains information from the seismological and structural engineering communities to reduce the risk of damage from earthquakes.

A minimum of two exits from the space, and exit signs throughout are required. A sprinkler system is installed for fire suppression.

Facilities Required

Storefront, Better Sportswear, Dressy Sportswear, Moderate Separates, Casual Knits, Jeans and Sweats, Men's, Accessories, Cash Wrap, Bar, Sitting Area, and Fitting Rooms. Non-selling areas include Restroom, Office, Stockroom, and rear exiting corridor.

Spatial Requirements

STOREFRONT AND SHOW WINDOW

- Review the Tenant Design Manual and adhere to the preliminary submittal requirements as outlined in Phase II of the submittal requirements. Quality materials must be used and must be well detailed to bring out their best characteristics. Creativity and originality are essential, and the landlord reserves the right to reject any storefront that does not exhibit these qualities.
- Plan and design the entrance, storefront, show window, and signage elements. Limit design elements to the allowable storefront area indicated on the elevation in Figure 9–4. The mall trim, base, and column facing are manufactured limestone. The mall floor finish is black granite. Tenants are encouraged to create a show window design for their storefronts, with a distinct entrance and large display windows. Swinging doors must not swing across the tenant's lease line. Roll-down doors are not allowed.

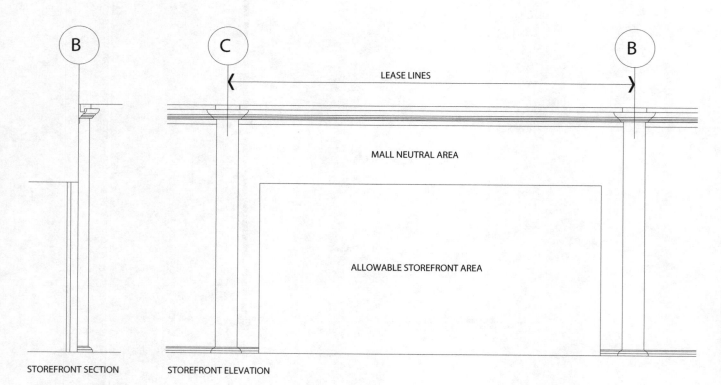

STOREFRONT SECTION   STOREFRONT ELEVATION

Figure 9–4   Storefront Elevation and Section

- The sign must be designed as an integral part of the storefront and be located at least 7 feet above the finished floor (to the underside of the sign). The storefront signage is to incorporate the same type style indicated on the business card shown in Figure 9–1.
- Show window interior will be approximately 30 square feet. Provide an elevated platform area to display merchandise with view into the interior of the store. Use theatrical lighting fixtures. Provide power for three to five small TV monitors with remote VCR connections.
- The mall will provide a drywall bulkhead above the allowable storefront area. It will be painted drywall with a light cove. Also provided by the mall is the common area flooring up to the lease line.

BETTER SPORTSWEAR

- 250 square feet. Adjacency: Moderate Sportswear
- Sweaters and sweater dresses: 14 units[2] folded
- Knit skirts and sweaters: 34 units hanging and folded
- Linen separates: 31 units hanging
- Skirts: 22 units hanging
- Sweaters: 40 units folded
- Shirts and tops: 14 units hanging
- Jackets: 14 units hanging
- Pants: 4 units hanging
- Dresses: 6 units hanging
- Suits: 3 units hanging
- Coats: 3 units hanging

DRESSY SPORTSWEAR

- 50 square feet. Adjacency: Better Sportswear
- Skirts: 8 units hanging
- Blouses: 8 units hanging
- Vests: 8 units hanging

MODERATE SEPARATES

- 300 square feet. Adjacency: Front of store
- Sweaters: 60 units folded
- Shirts: 64 units hanging and folded
- Jacket, skirt, and pant coordinates: 72 units hanging
- Skirts: 15 units hanging
- Jacket, skirt, top coordinates: 50 units hanging
- Tunics, tuxedo shirts, and pants: 28 units hanging
- Tunics and skirts: 28 units hanging

CASUAL KNITS

- 300 square feet. Adjacency: Jeans and Sweats
- Dresses, skirts, pants, and tops: 43 units hanging
- Jackets and pants: 58 units hanging
- Jackets, skirts, and tops: 68 units hanging
- Jackets, skirts, pants, and tops: 80 units hanging
- Dresses: 20 units hanging

JEANS AND SWEATS

- 50 square feet. Adjacency: Men's

---

2. Units are the number of garments or merchandise items, also referred to as the SKU (store keeping units).

- Jeans: 35 units hanging and folded
- Jean jackets: 18 units hanging
- Skirts: 8 units hanging
- T-shirts: 10 units folded
- Tango sweats: 24 units hanging

MEN'S

- 200 square feet. Adjacency: Jeans and Sweats
- Basco jackets and pants: 101 units hanging
- Basco shirts: 22 units folded
- Jackets: 34 units hanging
- Pants: 12 units hanging
- Jackets, pants, and shirts: 54 units hanging
- Sweaters: 12 units folded
- Shirts: 26 units hanging
- Knit shirts: 12 units folded
- Shorts: 12 units folded
- T-shirts: 12 units folded
- Memphis neckwear: 32 units folded
- Sneakers: 6 pair

ACCESSORIES

- 35 square feet. Adjacency: Cash Wrap, if possible
- Merchandise includes belts, jewelry, beads, scarves, and beach bags.
- Include impulse item sleepwear consisting of packaged robes, slippers, and pajamas: 61 units folded.

CASH WRAP

- 40 square feet (approximate). Adjacency: Accessories
- It should be centrally located. It is important that the salespeople are able to see the entrance and the fitting rooms from the Cash Wrap area. ADA requires that cash wraps be on an accessible route and double cash wraps are required to be 36″ apart.
- The area is used to write up sales and wrap the merchandise. Provide storage for sales receipts and miscellaneous working supplies and a locking cash drawer. Storage is also required for bags, tissue, and plastic garment bags. Bag sizes: plastic garment bags are 18″ wide (roll), shopping bags are 8½″ × 4 ⁵⁄₁₆″ × 9½″ and 16″ × 6″ × 15½″. Tissue size is 20″ × 30″.
- Electrical equipment in this area will be a computerized cash register, calculator/adding machine, credit card and credit verification machine, telephone, stereo equipment, and three to five VCRs with remote connections. The televisions sets will be located at the bar or show window.

BAR

- 90 square feet. Adjacency: Sitting Area
- They will be serving complimentary wine, champagne, soft drinks, coffee, and small tea sandwiches.
- Provide seating for three.
- Plan an area for three to five small TV monitors with remote VCR connections.
- Storage is required for plates and serving ware, flatware, glassware, liquor, coffee, and coffee machine.
- Provide an under counter refrigerator. No sink is required.

SITTING AREA

- 90 square feet. Adjacency: Accessories. Secondary: Bar
- Consider the bar and seating areas as places to mingle and chat, read magazines, or wait for a consort. Marc would like her customers to feel comfortable enough so that they can open up and get themselves in a better mood.
- Provide a seating group and a large coffee table for magazines.

FITTING ROOMS

- Three fitting rooms (4′ × 6′ minimum) are required and one is to be handicap accessible.
- Provide a mirror, small seat, shelf, and garment hooks in each fitting room.

RESTROOM

- 36 square feet
- The restroom is primarily for employees and customers. There are additional handicap facilities in the mall. Verify the local codes and ADA guidelines to determine necessary compliance to accessibility regulations for a shop of this size.
- Storage is required for employee belongings and cleaning materials.
- Provide one water closet and a hand-washing sink.

OFFICE

- 50 square feet minimum. Adjacency: Stockroom. The office must have access to and from the stockroom.
- Provide two separate work areas for the owner and the manager. Each work area has to have a work surface, chair, shelving, and file storage. Provide additional storage units.
- The safe will be located in this room and its size is 36″ wide × 24″ deep × 42″ high.
- Telephone, computer, fax, and printer equipment will be used.

STOCKROOM

- 180 square feet. Adjacency: Office and rear exit door
- This space is used for back stock of merchandise, receiving, steaming, and ticketing garments.
- Provide a double hang bar and shelving. Provide a small work counter and a small, portable table with stacking or folding chairs for employees.

SUBMITTAL REQUIREMENTS

Phase I: Programming

Part A Due: _____   Part B Due: _____   Part C Due: _____

A. Research various codes, arrangements, and clearances for the above-outlined spaces. Research fixture, furniture, and equipment for the same. Research the merchandise classifications, vendors, and designer lines planned for the store. Analyze the program outlined above and develop a programming matrix keyed for adjacency locations. A sample blank matrix is provided in Chapter 31.

B. With the information gathered from the matrix, provide a minimum of three bubble diagrams for the spatial relationships of the areas required.

C. Provide a one-page typed concept statement of your proposed planning direction and initial design concept.

Presentation formats for Parts A through C: White bond paper, 8½″ × 11″, with a border. Insert in a binder or folder. The matrix and bubble diagrams are to be completed on the CADD program.

Evaluation Document: Programming Evaluation Form

Phase II: Conceptual Design

Part A Due: _____   Part B Due: _____

A. Develop a preliminary partition and department location plan showing all walls, ceilings, built-in fixtures, fitting rooms, doors, and windows. Label departments and rooms and indicate the square footage required and the square footage planned. These plans can be freehand drawings or CADD drawings in 1/4″ scale. Drawings must have an appropriate border. Finished drawings must be a blueprint or white bond copy.

B. Develop the Storefront and submit as outlined in the Tenant Design Criteria Manual:

Submission I—Preliminary Design Phase

Tenant's Designer shall submit one (1) set of sepia transparent drawings and one (1) set of colored or rendered prints and/or photographs of store design along with color samples of proposed materials relating to the storefront and Criteria Control Area to the Landlord's Tenant Coordinator for preliminary review and approval.

These preliminary designs are to be submitted within twenty-one (21) calendar days or less of signing the Lease, subject to opening commitment as provided in the lease and shall consist of at least the following drawings, clearly identified with the Project name, Tenant's store name, Tenant leased premises number and floor plan, (Exhibit "B" of Lease) with Tenant leased premises clearly identified. Note: Only drawings on a standard size of 24″ × 36″ will be accepted.

a. Floor plan of the premises, showing approximate store fixture locations, interior partitions, toilet rooms, exits, display cases, etc. (Scale: 1/4″ = 1′0″)

b. Reflected ceiling plan showing the type of ceiling and finish; and the type of lighting fixture and locations. (Scale: 1/4″ = 1′0″)

c. Storefront elevation(s) showing signage, display concepts, and closure. Identify all materials to be used. (Scale: 1/2″ = 1′0″).

d. Large scale section(s) through the storefront (Scale: 1″ = 1′0″)

e. Sketches, perspectives, sections, or other details that will clarify the design of the storefront and the Criteria Control Area.

Evaluation Document: Conceptual Design Evaluation Form

Phase III: Design Development

Part A Due: _____   Part B Due: _____   Part C Due: _____

A. Develop the above plans into a final floor plan, showing all items noted in the program. This plan is to be completed using the CADD program. Develop ceiling and lighting studies into a final reflected ceiling plan, showing all ceiling conditions and heights on the CADD program.

B. Develop a color, material, and final fixture scheme. Samples and pictures can be presented in a loose format.

C. Develop a perspective sketch of the interior of the store, showing all design elements and furnishings. Refine the storefront elevation and sections, showing all design and signage elements.

Evaluation Document: Design Development Evaluation Form

Phase IV: Final Design Presentation   Date Due: _____

A. The final presentation will be an oral presentation to a panel jury. Project must be submitted on 20″ × 30″ illustration boards. Your name, the date, the project name and number, the instructor's name, and course title must be printed on the back of each board in the lower left-hand corner. Submit as many boards as required, including a colored rendering, storefront elevation, fixture and furniture selections, material selections, and plans.

The plans should be presented in 1/4″ scale. The elevations are to be presented in 1/2″ scale.

Evaluation Document: Final Design Presentation Evaluation Form

Phase V: Construction Documentation   Due Date: _____

A. Prepare interior working drawings and specifications for non-load-bearing interior construction, materials, finishes, furnishings, fixtures, and equipment. The drawings and specifications must include the following as outlined in the Tenant Design Criteria Manual:

Drawing Submission Criteria and Approval

A. Interior Drawings and Specifications

    (1) Floor Plan (indicating . . .)
        (a) Lease line relationship to stores
        (b) Landlord's and/or Tenant's responsibilities
        (c) Location of partitions, doors, and proposed fixtures
        (d) Overall dimensions of space and column locations and column lines

    (2) Reflected Ceiling Plan (indicating . . .)
        (a) Layout and specification of ceiling system
        (b) Ceiling heights for all areas
        (c) Light fixtures, schedule and description, manufacturer's name, model number, and type
        (d) Supply and return grilles
        (e) Fire sprinkler head locations
        (f) Other items attached to or coming through ceiling

    (3) Elevation of Storefront (indicating . . .)
        (a) Materials used and color of materials
        (b) Type of signage and proposed locations
        (c) Section through storefront
        (d) Rendering perspective or photograph of similar storefront, if existing

    (4) Specifications
        (a) Specifications, if not on drawings, should be submitted on an 8½″ × 11″ format.

B. The drawings must also include the following as a typical bid package:

    (a) Any special design elements and attached interior fixtures keyed to details, sections, and elevations as required on the above plan

    (b) Elevations or details of interior walls showing design elements, fixtures, and indications of the colors and materials, keyed to details and finish specification schedules

(c) A floor covering plan and details keyed to specification schedules

(d) Details showing sections through specially designed elements or fixtures, which must be prepared and identified as to wall locations

(e) An electrical power criteria plan, which must be provided and keyed to a legend. Plans or elevations, including stereo speakers, security cameras, and emergency lighting and exit signs

(f) The reflected ceiling plan, including stereo speakers, security cameras, and emergency lighting and exit signs

(g) All materials and finishes, which must be specified on color and material schedules

(h) Title sheet, including index of drawings, project representatives, applicable governing codes, project statistics, utility service representatives, and title block

C. The plans should be presented in 1/4″ scale. The elevations are to be presented in 1/2″ scale.

Evaluation Document: Construction Documentation Evaluation Form

Project Time Management Schedule: Schedule of Activities, Chapter 31.

The AutoCAD LT 2002 drawing file name is tango.dwg and can be found on the CD.

REFERENCES

Books

Ballast, David Kent. 1994. *Interior construction and detailing for designers and architects.* Belmont, Calif.: Professional.

Israel, Lawrence J. 1994. *Store planning/design: History, theory, process.* New York: Wiley. (See p. 71 for a sample schedule.)

Sudjic, Deyan. 1990. *Rei Kawakubo and comme des Garcons.* New York: Rizzoli International.

Trade Magazines

Women's Wear Daily: http://www.wwd.com

Trade Organizations

See previous retail project.

Fixture Manufacturers

Spartan Showcase, Inc.: http://www.spartanshowcase.com. See previous retail project.

Special Resources

Fixture dimension criteria are provided in Chapter 31.

Project code reference: The City of Newport Beach, Building Department: http://www.city.newport-beach.ca.us/building

CHAPTER 10

Venetia's City Boutique

OUTCOME SUMMARY

The objective of this exercise is to develop your proficiency to interpret and incorporate program information into bubble diagrams and a block plan solution within a three-hour time limit. It is a Charette, which will provide experience in working within a short deadline. The project then continues to include developing conceptual skills through schematic or initial design concepts and detailing and refining ideas from the schematic design phase within a customary time schedule.

client profile

Venetia and Gary Tivoli are the owners of a women's specialty store called Venetia's Boutique. The merchandise is upscale, youthful, and highly fashionable, and includes Laurel, Bisou-Bisou, BCBG, Votre Nom, Renfrew, ABS by Allen Schwartz, Joop, Nicole Miller, and Apriori, to name a few. Venetia fills the boutique with fashions that have an edgy modern style and a bit of the flirtatious, too. The owners plan to open a second location at One Oxford Centre in the downtown area, and it is to be called Venetia's City Boutique.

You have designed three other stores for Venetia and this time the dare of coming up with a fresh and unusual design increases for the city boutique. Simplicity has always reigned in the design approaches for the other stores. You and Venetia are of the same opinion that a clean and minimal approach in a loft-like setting would be a good match for the urban store. Venetia has stressed the importance of approachability: that she does not want the store to be intimidating or over-sophisticated. For Venetia, selling is an art form, and the fitting rooms are an important stage of the process. She has a great talent for finding the perfect garment for the customer. A considerable amount of networking and socializing are part of both the shopping and fitting experience. She will be leasing a small part of her boutique to Caesar, an emerging custom jewelry designer with a great flair for clean minimalism in his creations.

One of the opening promotions for the store, as shown in Figure 10–1, is called "Undress for Success," developed by Elaine Restauri, publicist and graphic designer. And, of course, Gary will have the "Harley" at hand for a second time to show off the one-of-a-kind Nicole Miller design for the bike. Move over and make some room because he just may want to bring a Reynard Indy car in once again to use as a tie rack! Other types of promotions throughout the year include trunk shows, Nicole Miller bridal shows, fashion shows, and charitable community events.

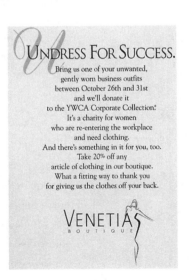

UNDRESS FOR SUCCESS.

Bring us one of your unwanted, gently worn business outfits between October 26th and 31st and we'll donate it to the YWCA Corporate Collection. It's a charity for women who are re-entering the workplace and need clothing. And there's something in it for you, too. Take 20% off any article of clothing in our boutique. What a fitting way to thank you for giving us the clothes off your back.

VENETIA'S
BOUTIQUE

Figure 10–1 Opening promotion. Elaine Restauri, Graphic Designer.

The Shops at One Oxford Centre are owned and managed by Oxford Development Company. Designed by Hellmuth, Obata, Kassabaum Architects and built in 1983, the forty-story building features over 85,000 square feet of upscale specialty retail shops. Venetia's neighbors include Rodier, Emphatics, Kountz and Rider, Ann Taylor, Talbot's, Hardy and Hayes, St. Moritz Chocolats, Au Bon Pain, Philip Pelusi Hair Salon, Asiago, Dingbats, and the Rivers Club. Oxford provides on-site management for the tenants and 840 indoor parking spaces for tenants and customers.

Previously occupied by a men's clothing and shoe retailer, the store atmosphere was very masculine, dark, and outmoded. Lighting issues and an exposed plenum ceiling

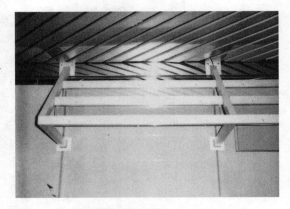

Figure 10–2 Partial Storefront and Mall Ceiling

space (painted black), heighten the planning difficulties, including the sharp triangular character of one end of the space. The fixtures are still in the space and also are available for use. The client wants you to use as many of the existing elements in the space as possible because of the minimal budget allowance for the new store. The numerous existing marble floor level changes and wall fixture configurations in the existing space would make the store a safety risk for the customers and would be very difficult to service, respectively. After your visit to the site you concluded that the existing interior architectural and fixture elements are unusable with the exception of the exposed ceiling (at 13′0″ above the finished floor) and open wood grid (9′10″ above the finished floor). You will, however, be able to use the existing showcase elements and incorporate them into the Cash Wrap and Accessories areas. They will also be needed for Caesar's jewelry display and work area. Considering the budget limitations, it was concluded that the bulk of money should be spent on lighting, new fitting rooms, and new floor fixtures.

There is a great span of glass storefront window totaling 72 linear feet to be addressed. It is your decision whether or not to reuse the existing metal awning at the storefront as shown in Figure 10–2 and Figure 10–3. It will be a challenge to provide several display groupings that will invite the customers inside. The owners do not want mannequins in this store because they are difficult to service without a display staff.

Figure 10–3
Storefront and Mall Ceiling

Venetia has expressed the desire for casters on the floor fixtures to make them easier to move, because flexibility is an important feature to incorporate in all floor fixture functions. Incorporate the ability to hang in a face-out configuration with shelving on a few specialty fixtures. The preference is to use straight hang bars on the perimeter walls. Mirrors are to be strategically located throughout the store.

project details

General and Architectural Information

Location: Venetia's City Boutique, The Shops at One Oxford Centre. Lease Space #R.245, Retail level 2

Gross square footage: 1,327. Net square footage: 1,250

Hours of operation: Monday–Friday: 10 a.m. to 6 p.m.; Saturday: 10 a.m. to 5 p.m.

Project budget: $75,000.00 (this budget is very low and would be more realistic at twice the amount). No bidding is required. The owners will act as the general contractor.

Opening or move-in date: Soft opening, October 1st; Grand opening, October 17th

Existing construction: Current facility plans and specifications are not available. Mall facility plans are available. Site information gathering is required.

Consultants required: Architect, electrical engineer, and mechanical engineer

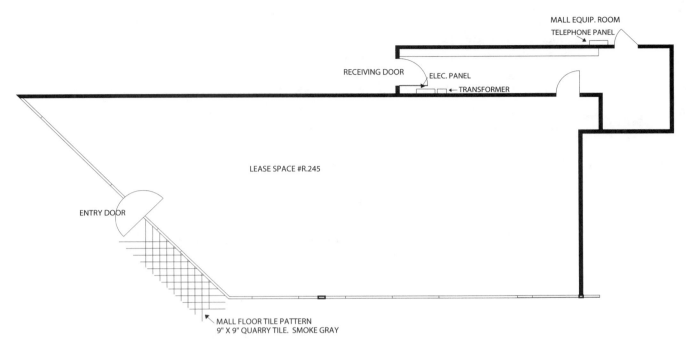

MALL EQUIP. ROOM
TELEPHONE PANEL

RECEIVING DOOR   ELEC. PANEL

TRANSFORMER

LEASE SPACE #R.245

ENTRY DOOR

MALL FLOOR TILE PATTERN
9" X 9" QUARRY TILE. SMOKE GRAY

Figure 10–4 Lease Plan

Security: Security cameras with closed circuit TV will be used, and Oxford provides 24-hour security. Sensormatic equipment is not required at the entrance. Lock devices are not required on fixtures, excluding the showcases and cash drawers. The storefront window interior lighting is to be on a timer system.

Codes and Governing Regulations

Occupancy: M, Mercantile

Building type: Type 2A. Protected

Design the project to comply with the most current edition of the BOCA National Building Code, the National Electrical Code (NEC), and the Americans with Disabilities Act.

A minimum of two exits from the space and exit signs throughout are required.

Since the completion of this project, the Pennsylvania Department of Labor and Industry will adopt as its Uniform Construction Code (UCC) the International Building Code (IBC) and most of its referenced standards.[1]

Fire Protection: Fire Sprinkler Suppression System

Facilities Required

Sales Floor: Nicole Miller, Casual Sportswear, Suits and Dresses, PM Dressing and Bridal, Accessories, Jewelry, Caesar's Designs, Cash Wrap, Sitting Area, Bar, and Fitting Rooms. **Back Areas:** Stockroom and Office

Spatial Requirements

SPORTSWEAR

This department is to be located at the entrance of the store, prominence being given to the Nicole Miller boutique classification. This section will be a combination of hanging and fold-

---

1. Commonwealth of Pennsylvania, Pennsylvania Department of Labor and Industry.

ed merchandise to be presented on both perimeter wall fixtures and floor fixtures. Merchandise classifications include:

- Nicole Miller Boutique: 360 to 450 units. The designers' collection includes the complete accessory line for women in addition to the dresses, suits, and separates categories. Also featured are the Nicole Miller celebrated silk print ties and boxer shorts for men. The designer created a special Pittsburgh print exclusively for Venetia's boutiques. A storefront show window display area is to be dedicated to Nicole Miller merchandise.
- Casual Sportswear: 200 to 300 units hanging
- Jeans: 50 to 75 units hanging and folded
- Sweaters: 100 units folded and/or hanging
- Separates: 50 to 100 units folded or hanging
- Collections

  This is the featured and high-end merchandise to be presented as individual groupings. Locate this classification adjacent to Suits and Dresses.

  - Laurel

    Group 1: 51 units hanging, 12 units folded

    Group 2: 43 units hanging

    Group 3: 40 units hanging

  - Apriori

    Group 1: 24 units hanging

    Group 2: 19 units hanging

    Group 3: 29 units hanging

    Group 4: 29 units hanging

SUITS AND DRESSES

- Suits: 25 to 50 units hanging
- Dresses: 25 to 50 units hanging

PM DRESSES AND BRIDAL

- Evening dresses: 40 units hanging. Gowns need to be hung 63″ from the top of the hang bar in order to clear the floor.
- Bridesmaid dresses: 40 units hanging. These are Nicole Miller sample dresses that will be custom ordered.

ACCESSORIES

Locate near Cash Wrap area and provide a display for some of the items throughout the store. Merchandise includes:

- Small handbags and hats: 50 to100 units. Shelving is required.
- Scarves: 20 to 30 units. Folded merchandise, is to be displayed with handbags, jewelry, or garments.
- Belts: 100 units. Hanging merchandise, must be on special fixture with short pegs.

JEWELRY

- Costume jewelry: 250 to 300 units. Adjacency: Cash Wrap
- All merchandise must be displayed in showcases. Provide custom case interior and jewelry display platforms.

Figure 10–5 Exterior of existing showcases

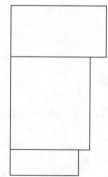

SHOWCASE SECTION

Figure 10–6 Showcase profile drawing

CAESAR'S DESIGNS

- Adjacency: Cash Wrap
- Provide two showcases for the display of 75 rings, 12 pendants, 12 pairs of earrings and 3 sets that each include a pendant and earrings. Existing storage in showcases is adequate for these requirements.
- Merchandise is to be displayed in showcases. Provide custom case interior and jewelry display platforms.
- A work surface is required with proper task lighting.
- Provide two customer stools, a work stool, and countertop mirrors.

CASH WRAP

- It is important that the salespeople are able to see the shop entrance and the fitting rooms from the Cash Wrap area. Locate it so there are no blind spots from this viewpoint.
- The area is used to write up sales and wrap the merchandise. Storage is required for bags, tissue, and plastic garment bags. Electrical equipment in this area will be a calculator/adding machine, credit card terminal, and telephone. A computer will not be used in this area. The cash wrap configuration at the Galleria store is too large. See Figure 10–7.
- You are required to use the existing showcase/wrap fixture components as shown in Figure 10–5 and Figure 10–6. The drawers in these fixtures are adequate for tissue, bags, and miscellaneous items, and there are two locking drawers for cash. You can subtract from, add to, refinish, and reconfigure these elements. There are a total of four showcases: two are 5'0" × 1'10" and the other two are 4'0" × 1'10".
- Holds and alterations are to be located in this area and will require a hang bar of 2 linear feet.

Figure 10–7
Cash wrap. Venetia's Boutique, Galleria. Bill Paveletz, Photography.

SITTING AREA

- This seating area is to be a place to gather and get in touch with friends, not necessarily to buy. Venetia and Gary understand the need to simply decompress and alter the mood, and they want their customers to feel that it can be done here. They will offer music, videos, soft drinks, and wine. Waiting companions will frequently occupy the space, and Gary has requested a TV or two for watching fashion shows or sports programming.
- A sofa and table furniture grouping or single chairs should be planned.

BAR

- This should not be an obvious looking bar setup and requires a maximum of 2 or 3 linear feet. No seating is required.
- Provide storage for glasses, ice bucket, coffee cups, coffee maker, wine, and soft drinks.
- A small under counter refrigerator will be needed.

FITTING ROOMS

- Two fitting rooms are required and one is to be handicap accessible.
- Provide a mirror, a comfortable chair, a shelf, garment hooks, and a hang bar in each fitting room.
- Venetia especially likes the use of drapery for the fitting room area because it gives her the ability to flow back and forth from the sales floor to the customer.
- Area between the two fitting rooms will function as a fitting area and must have some visibility to the sales floor. It needs a large three-way mirror, fitting platform, a shelf for pins, and numerous bars or hooks on the walls for merchandise that is being tried on. The three-

way mirror may be at a permanent angle or of a hinged panel construction. Research proper angle of mirror.

STOCKROOM

- This space is used for a minimum back stock of merchandise and receiving, steaming, and ticketing garments.
- Provide a double hang bar and shelving. If space allows, provide a small work counter and chair.
- The area exists and is located near the receiving door. The shelving can be reused.

OFFICE

- The Office must be located off the sales floor, in the back area adjacent to the Stockroom. It does not need to be an enclosed room.
- Provide work surfaces for two people, file storage, and shelving. The safe is 24″ wide × 24″ deep × 24″ high.
- One computer and two telephones are required.
- The VCR, stereo system, and security TV monitor are to be located in this area.

SUBMITTAL REQUIREMENTS

Phase I: Programming   Due at end of three-hour time limit.

A. Analyze the program outlined above, the existing conditions and limitations, and develop a programming matrix keyed for adjacency locations.

B. With the information gathered from the matrix, provide a minimum of three bubble diagrams for the spatial relationships of the areas required, excluding the non-selling areas (Stockroom and Office). Use the scaled (1/8″ = 1′0″) floor plan in Figure 10–9 and sketch your ideas on tissue overlays.

C. Develop a block plan indicating department locations, showing all walls, Fitting Rooms, showcases and Cash Wrap, doors, and windows. Label departments and rooms and indicate the square footage required and the square footage planned. These plans can be freehand drawings. Drawings must have an appropriate border.

D. At the end of the three-hour time limit, present your solution to the class. Your verbal presentation must include justification for your spatial and functional solutions, as well as any visual ideas that express the character of the space.

Evaluation Document: Programming Evaluation Form

MATRIX

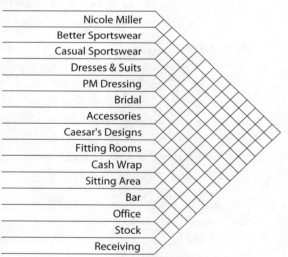

Figure 10–8   Matrix

LEVEL OF INTERFACE, COMMUNICATION OR ADJACENCY

- PRIMARY
- SECONDARY
- MINIMUM TO NONE
- UNDESIRABLE

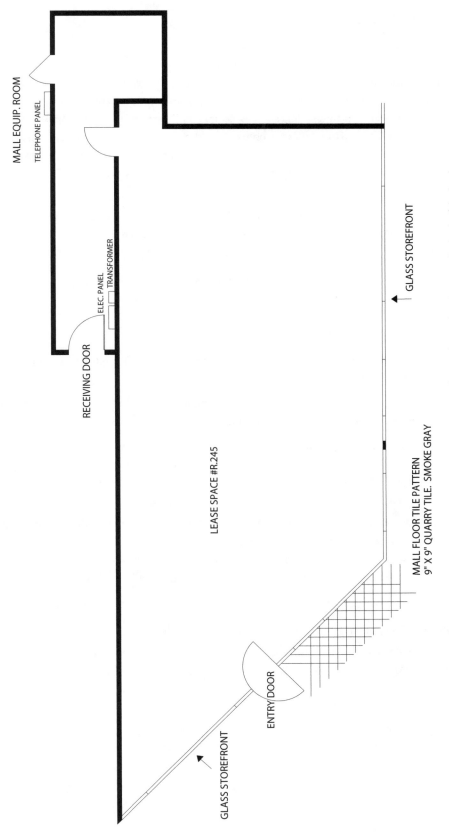

MALL EQUIP. ROOM

TELEPHONE PANEL

ELEC. PANEL

TRANSFORMER

RECEIVING DOOR

LEASE SPACE #R.245

GLASS STOREFRONT

MALL FLOOR TILE PATTERN
9" X 9" QUARRY TILE. SMOKE GRAY

ENTRY DOOR

GLASS STOREFRONT

Figure 10-9   Floor Plan. Print the floor plan from the AutoCAD drawing located on the CD in the back of this book.

Phase II: Conceptual Design

Part A Due: _____   Part B Due: _____

A. Research various codes, arrangements, and clearances for the above-outlined spaces. Research fixture, furniture, and equipment for the same. Research the merchandise classifications, vendors, and designer lines planned for the store. Develop a preliminary partition and department location plan, showing all walls, ceilings, built-in fixtures, fitting rooms, doors, and windows. Label departments and rooms and indicate the square footage required and the square footage planned. These plans can be freehand drawings or CADD drawings in 1/4″ scale. Drawings must have an appropriate border. Finished drawings must be a blueprint or white bond copy.

B. Develop the above plans into a final floor plan, showing all items noted in the program. This plan is to be completed using the CADD program. Develop ceiling and lighting studies into a final reflected ceiling plan, showing all ceiling conditions and heights on the CADD program.

Evaluation Document: Conceptual Design Evaluation Form

Phase III: Development

Part A Due: _____   Part B Due: _____

A. Develop a color, material, and final fixture scheme. Samples and pictures can be presented in a loose format.

B. Develop a perspective sketch of the interior and an elevation of the storefront entrance, showing all design elements and furnishings.

Evaluation Document: Design Development Evaluation Form

Phase IV: Final Design Presentation   Date Due: _____

A. The final presentation will be an oral presentation to a panel jury. Project must be submitted on 20″ × 30″ illustration boards. Your name, the date, the project name and number, the instructor's name, and the course title must be printed on the back of each board in the lower left-hand corner. Submit as many boards as required, including a colored rendering, colored elevation, fixture and furniture selections, material selections, and plans.

The plans should be presented in 1/4″ scale. The elevations are to be presented in 1/2″ scale.

Evaluation Document: Final Design Presentation Evaluation Form
Project Time Management Schedule: Schedule of Activities, Chapter 31.
The AutoCAD LT 2002 drawing file name is venetia.dwg and can be found on the CD.

REFERENCES

Trade Magazines

See previous retail projects.

Trade Organizations

See previous retail projects.

Fixture Manufacturers

See previous retail projects.

Special Resources

Fixture dimension criteria are provided in Chapter 31.

Project reference:

ABS: http://www.absstyle.com
BCBG: http://www.bcbg.com
Nicole Miller: http://www.nicolemiller.com

CHAPTER 11

Ultima Moda

OUTCOME SUMMARY

The project features three phases of the design process: identifying and analyzing the client's needs and goals, developing conceptual skills through schematic or initial design concepts, and detailing and refining ideas from the schematic design phase. The project focuses on exposure to working with existing architectural conditions and existing fixtures for a retail environment, thus increasing proficiency in problem-solving skills for complex and challenging issues.

client profile

Figure 11–1 Logo

Hal, the California men's retailer, became well known for the quality merchandise that has been delighting his customers for four decades. Ultima Moda's collection of men's high-end, urban clothing sells well because of the retailer's service as well as the quality of the merchandise. Most of the buying for the store merchandise is done in Italy. Ermenegildo Zegna, Salvatore J. Cesarini, and many other Italian clothing lines are found here along with other favorites such as Jhane Barnes and Perry Ellis.

In 1986 you teamed with Hal and his sons Steve and Gary to renovate the southern California fifties landmark because the merchandise unmistakably outshone the décor, and today, history is repeating itself. Hal's Ultima Moda is relocating from the stand-alone store on the boulevard to Westwood Place. They had never occupied a space in a mall setting but it fits in their desire to increase their customers beyond the already loyal following. Westwood Place is a three–level outdoor shopping center with retail shops, retail services, dining, and office facilities. Specialty retailers include BCBG, Bo, M. Fredric, Chico's, Kiwi, Trio, Viavi, Renaissance Kids, En Ciel, D'Spa, Allen Edwards Salon, and 2.2.4.

In this store the client wants to capture a new look that will appeal to a younger generation while not alienating the older, established customers—all on a modest $150,000.00 budget. The only design preference is to continue with a clean and modern style that incorporates a faux painted wall texture.

The existing floor fixtures are to be refinished to work with the new design scheme and relocated to the new store. Several new feature fixtures can be added and, of course, the perimeter wall fixtures will be new. The budget does not allow for a completely new ceiling; therefore, you are to work with the existing ceiling conditions and lighting in the front portion of the space. It is possible to remove, relocate, and supplement the existing lighting fixtures. And to make your task even more difficult, the budget will not permit the removal of the present wood floor portion, so you must incorporate it into your scheme. Finally, the existing chairs and benches are to be reupholstered and incorporated into the new store.

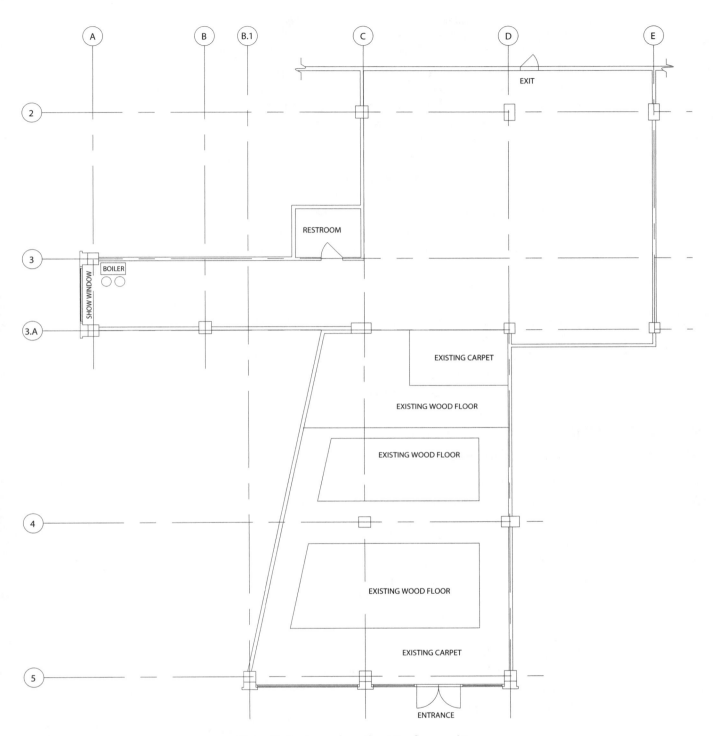

Figure 11–2 Lease plan with existing floor conditions

Two different lease spaces were available: 3,300 square feet and 4,300 square feet. Their existing store was 5,400 square feet so they had to reduce both the clothing and sportswear departments by at least 25%. Both lease plan options had show windows and signage facing the Boulevard; however, neither were optimal floor plans. Pondering their options, Steve and Gary decided that their answer might be twofold by using sections of each lease plan. The lease plan chosen, Figure 11–2, incorporates portions of each space. Both the plan and the budget restraints initiate a difficult design evaluation as you are beginning to see.

Your challenge is to navigate a somewhat complex lease plan outline, the irregular characteristics of the existing architectural elements, and the use of existing fixtures. The lease plan is, in effect, split in three parts. The front segment was a women's specialty shop with varying floor finishes and recessed ceiling coves above; the rear and side areas were previously unoccupied. Clarity is your goal regarding aisle flow in directing the patrons to each merchandise classification.

In the Sportswear Department, you have been asked to create segmented areas on the walls to vignette groups of collections. This vignette should be a combination of shelving, hang rods, face-outs, and displayed feature merchandise. The collections will include changeable groupings of shirts, sweaters, knits, slacks, jackets, sport coats, and jeans. They use simple display tables (item M-1 in Table 11–1, Fixture Survey) extensively throughout the store, and one should be located in front of each collection area wherever possible. These tables are used to feature merchandise and to assemble the customer's choices for final selection. In the Sportswear Department, they want to use tables (item M-1 in Table 11–1) with shelves underneath and no shirt cube floor fixtures.

The Clothing Department is to have perimeter fixtures in the typical double hang bar function for suits and jackets. Provide some visual relief to the strong horizontal lines created by this fixture orientation. Slacks are presented in the floor fixtures (item M-4, in Table 11–1) without exception. The Dress Shirt classification is to be located in an individual section and presented using a combination of wall shelving with table floor fixtures (item M-1, in Table 11–1) in front of the wall. These tables are to have shelving underneath. Mirrors are required along the walls for trying on jackets, and pullout hang bars are to be located at these positions.

project details

General and Architectural Information

Location: Hal's Ultima Moda, Fashion for Men, Suite 105, Westwood Place

Gross square footage: 3,596. Net usable square footage: 3,380

Initial number of employees in facility: Four salespeople, one tailor

Hours of operation: Monday–Friday: 10 a.m. to 9 p.m.; Saturday: 10 a.m. to 8 p.m.

Project budget: $150,000.00 for fixtures and accent lighting. Landlord will perform the general construction, and the owner will administer the bidding and construction of the interior fixture and finish work. The landlord will build out the space according to the tenant's specifications, subject to a mutually agreed-upon space plan. It will include, but not be limited to, the following:

- Finished walls with appropriate electrical
- Drywall ceiling with recessed lighting to tenant's specifications
- One room for the tailor and facilities for the boiler
- One stockroom
- One office
- Wall fixtures, including recessed standards, drywall, and appropriate soffits to house lighting as required
- Four dressing rooms
- In accordance with tenant's specifications, combination of new carpet and existing wood floor throughout
- Recessed, double door entry from the courtyard into the retail space

Figure 11–3 Exterior photo of the left side of the storefront as viewed from the boulevard

- One bathroom for handicapped persons
- Paint throughout in a color of tenant's choice

Tenant design criteria: Storefront architectural and design elements cannot be altered with the exception of locating the new entry doors. The exterior sign is provided by the landlord and is a standard design for all shops that allows the use of the shop logo or typestyle. The standard sign construction type is best illustrated in Figure 11–5.

Opening or move-in date: February

Existing construction: Current facility plans and specifications are not available. Mall facility plans are available. Site information gathering is required.

Consultants required: Architect, electrical engineer, and mechanical engineer. The landlord has the consultants on staff.

Security: Security cameras are required. Sensormatic equipment is not required at the entrance. Lock devices are not required on fixtures, excluding showcases and cash wraps.

Codes and Governing Regulations

Occupancy: Mercantile

Building type: Type II. One Hour

Design the project to comply with the most current edition of the ICBO Uniform Building Code (UBC), the National Electrical Code (NEC), the Uniform Fire Code (UFC), the California Building Code (CBC), and the Americans with Disabilities Act. Special seismic life safety issues apply to construction because the structure is in a risk level 4 seismic zone. The UBC contains information from the seismological and structural engineering communities to reduce the risk of damage from earthquakes.

A minimum of two exits from the space and exit signs throughout is required. A restroom facility and fitting room for disabled persons is required. A sprinkler system is not installed for fire suppression.

Facilities Required

Sales Floor: Sportswear, Clothing, Formalwear, Alterations, Accessories, Cash Wrap, and Fitting Rooms. **Back Areas:** Stockroom(s), Tailor, Office, Restroom for Handicapped Persons, and Display Windows

Figure 11–4   Exterior photo of the right side of the storefront as viewed from the courtyard entrance. A color copy of this photograph is located on your CD and the instructor's presentation CD.

Figure 11–5
Back show window as viewed from the boulevard

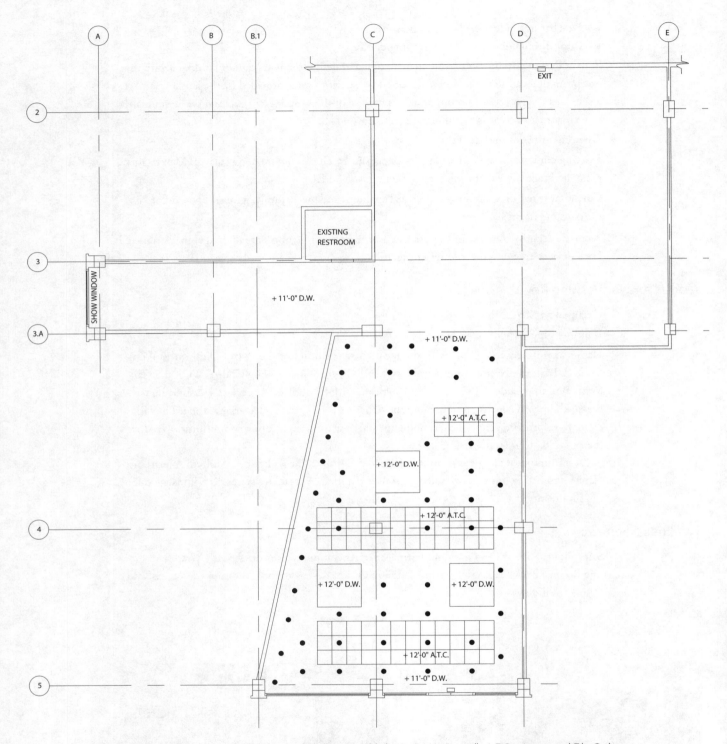

Figure 11–6 Existing ceiling conditions, including recessed lighting. D.W.: Drywall. A.T.C.: Accoustical Tile Ceiling.

Spatial Requirements

SPORTSWEAR

Adjacency: Locate this department at the entrance. The Cash Wrap and Seating areas must border this department. Lead with Casual Sport Coats and Dressy Sport Slacks.

Merchandise classifications include:

- **Sweaters and Knits:** Folded merchandise, must be located on perimeter wall fixtures. 150 to 175 units, 5 high (indicates the number of merchandise units that will be stacked together). There will be a total of 48 linear feet.
- **Sport Shirts:** Folded merchandise, must be located on perimeter wall fixture or floor fixture. 400 units, 4 high. There will be a total of 140 linear feet.
- **Casual Sport Coats:** Hanging merchandise, must be located on perimeter wall fixture and floor fixture. Merchandise must be presented on face-out hardware. 40 to 50 units. There will be a total of 10 to 15 linear feet.
- **Casual Cotton Slacks:** Folded merchandise, must be located on perimeter wall fixture and floor fixture. 150 units, 9 high. There will be a total of 25 linear feet.
- **Dressy Sport Slacks:** Hanging merchandise, must be located on perimeter wall fixture and floor fixture. Merchandise must be presented on face-out hardware. 120 units. There will be a total of 20 linear feet.
- **Jeans:** Folded merchandise, must be located on perimeter wall fixture or floor fixture. 50 to 60 units, 8 high. There will be a total of 10 to 12 linear feet.
- **Designer Jeans:** Hanging and folded merchandise, must be located on floor fixtures. 80 units.
- **Jackets:** Hanging merchandise, must be located on perimeter wall fixture and floor fixture. 50 units. There will be a total of 19 linear feet.
- **Jean Shirts:** Folded merchandise, must be located on perimeter wall fixture or floor fixture. 75 units, four high. There will be a total of 16 linear feet.
- **Knit Shirts and Tops:** Folded merchandise, must be located on perimeter wall fixture. 75 to 100 units. There will be a total of 40 linear feet.
- **Casual Belts:** Hanging merchandise, must be located on floor fixture. 125 units. There is an existing belt rack 27″ wide × 12″ deep × 57″ high.
- **Socks:** Folded merchandise, must be located on floor fixture with shelving. 250 units, 6 high.

CLOTHING

Adjacency: Locate at the rear of the store. Provide a seating arrangement in this department near the main fitting mirrors.

Merchandise Classifications include:

- **Suits:** Hanging merchandise, must be double hung on perimeter wall fixture. 200 units. There will be a total of 35 to 40 linear feet.
- **Sport Coats:** Hanging merchandise, must be double hung on perimeter wall fixture. 250 units. There will be a total of 45 linear feet.
- **Outerwear:** This includes topcoats and rainwear (short jackets are located in Sportswear). Hanging merchandise, must be single hung on perimeter wall fixture. There will be a total of 12 linear feet.
- **Feature Sport Coats:** Hanging merchandise, must be located on floor fixtures. Merchandise must be presented on face-out hardware. 20 to 25 units.
- **Dress Slacks:** Hanging merchandise, must be located on five existing floor fixtures (M-4). 400 units. They currently use 5 pant racks but it can be reduced to 4 pant racks.
- **Dress Shirts:** Folded merchandise, must be located on perimeter shelving wall fixture. 400 units, 4 high. There will be a total of 115 linear feet, 5 shelves high.

- **Ties:** Ties must be located on tie racks. 450 units.
- **Dress Belts:** Hanging merchandise, must be located on floor fixture. 125 Units. There is an existing belt rack 27″ wide × 12″ deep × 60″ high.

FORMALWEAR

- Dedicate a wall section within the Clothing Department for this merchandise.
- This includes tuxedos, dinner jackets, shirts, vests, and cummerbund and tie sets. The cufflinks and stud sets are to be located in the cash wrap showcase. Allow space for a display presentation shelf.
- **Tuxedo Jackets:** Hanging merchandise, 10 to 15 linear feet, must be located on perimeter wall. 25 to 30 units.
- **Shirts:** Folded merchandise, 10 to 15 linear feet of shelving, must be located on a perimeter shelving wall fixture above the jackets. 35 to 40 units.

ALTERATIONS

- 25-30 square feet. Adjacency: This area should be located on the sales floor, adjacent to the Fitting Rooms, Tailor, and Seating area if possible.
- Alterations functions as a fitting area and is considered to be a very strategic location that leads to the final sale.
- Provide a large fitting mirror with two side angles of 22.5 degrees.

ACCESSORIES

- Adjacency: Locate near Cash Wrap.
- The merchandise includes small gift items, small leather, handkerchiefs, and jewelry such as cuff links, tuxedo stud sets, tie bars, and collar stays.
- A small feature floor fixture and locking showcase in the Cash Wrap is suitable for this merchandise.

CASH WRAP

- Adjacency: Locate in a central area between Sportswear and Clothing.
- The area is used to write up sales and wrap the merchandise. Storage is required for bags, tissue, and plastic garment bags. Electrical equipment in this area will be the cash register terminal, calculator/adding machine, telephone, VCR, and stereo system.
- You are required to use the existing wrap fixture components. You can subtract from, add to, refinish, and reconfigure these elements. See Figure 11–7 and item M-9 in Table 11–1.
- A showcase is required at the Cash Wrap area for accessories.
- One or two televisions with cable and VCR connections are to be located in the store near the seating arrangements for occasional viewing of sports or tapes.

FITTING ROOMS

- Two are required for Clothing and one is required for Sportswear.
- Minimum size to be 4′6″ square. Preferred typical fitting room size is 4′0″ × 6′0″.
- One fitting room is to be handicap accessible and meet ADA guidelines (minimum size of 7′2″ × 7′6″).
- For security purposes, fitting rooms are not to be located near the rear fire exit or front entry doors.
- Each fitting room is to have a mirror, shelf, garment hooks, and a seat.

Figure 11–7
Cash Wrap Plan View and Interior Section Elevation

Stockroom(s)

- 165 square feet. Adjacency: Ideally, the stockroom is located near a receiving door but it is not a requirement in this case.
- This space is used for back stock of merchandise.
- Provide a double hang bar and shelving. If space allows, provide a small work counter and chair.

Tailor

- 200 square feet minimum. Adjacency: Alterations fitting area.
- This is the tailor's workroom where alterations are made to the clothing. A section of the space will be used for tagging and steaming garments.
- The tailor uses two sewing machines and they do not have to be located adjacent to each other.
- Reuse the existing fixtures and equipment. See Table 11–1, Fixture Survey.
- Sewing machines (F-13) and the blind stitch table (F-15) must be adjacent to a worktable (F-17). The pressing machine (F-12) must be located next to the boiler equipment. The ironing table (F-14) must be adjacent to pressing machine (F-12).
- Provide a garment hang bar above the tailor's work areas.
- The utility table (F-18) can be eliminated if necessary. The portable rack (F-19), normally located by the pressing machine, could be smaller or eliminated if necessary.
- The boiler equipment is located on the plan and cannot be moved.
- Any unused portions of the room must be used for stock and have shelving and hang bars.

Office

- Reuse existing furniture and equipment. See Table 11–1, Fixture Survey.
- Include a desk, a credenza, a desk chair, side chairs, a recliner, a refrigerator, and a microwave.
- The safe is to be located in this space.
- Equipment required are a computer, a telephone, a fax machine, and a copy machine.

Restroom for Handicapped Persons

- Existing handicapped accessible restroom will be reused and is located on the plan.
- Do not provide access directly from the sales floor.

Display Windows

- Enclosed Display Window is located on the plan. Design a special wall treatment and lighting elements.
- Storefront windows at the front entrance should not have enclosed display windows. Design display elements for the window section between the columns.
- You must reuse existing fixtures wherever possible.

ITEM NO	QUANT	DESCRIPTION	DIMENSIONS WIDTH × DEPTH × HEIGHT (INCHES)
Fixtures			
M-1	9	Clear glass (½") top. Black gloss plastic laminate base	62 × 30 × 30
M-2	3	Clear glass (½") top. Black gloss plastic laminate base	46 × 27 × 28
M-3	4	Clear glass (½") top. Black gloss plastic laminate base	36 × 36 × 30
M-4	5	Clear glass (½") top. Black gloss plastic laminate base	50 × 33 × 59
M-6	1	Clear glass (½") top. Black gloss plastic laminate base	30 × 30 × 18
M-7	2	Clear glass (½") top. Black gloss plastic laminate base	18 × 18 × 24
M-8	2	Clear glass (½") top. Black gloss plastic laminate base. 2 glass shelves under top	61 × 31 × 38
M-9	1	Cash wrap. Clear glass showcase. Black gloss plastic laminate. Mirror base	93 × 110 × 42 overall (See Fig. 11-7)
Furniture			
F-1	2	Upholstered chair	35 × 27 × 28 (back height)
F-2	5	Upholstered bench	30 × 19 × 21 (arm height)
F-3	1	Metal safe	20 × 24 × 30
F-5	1	Metal desk	72 × 36 × 29
F-6	1	Metal credenza	72 × 19 × 30
F-7	1	Metal file cabinet	18 × 26 × 52
F-8	1	Upholstered desk chair	23 × 27 × 42
F-9	1	Recliner	32 × 36 × 42
F-10	1	Refrigerator	28 × 24 × 63
F-11	1	Microwave	26 × 16 × 16
Tailor			
F-12	1	Pressing machine	60 × 42 × 63
F-13	2	Sewing machine	48 × 20 × 30
F-14	1	Ironing table	72 × 31 × 30
F-15	1	Blind stitch machine with table	48 × 20 × 30
F-16	1	Worktable	72 × 43 × 31
F-17	1	Worktable	72 × 31 × 31
F-18	1	Utility table	60 × 34 × 30

Table 11-1 Fixture Survey

SUBMITTAL REQUIREMENTS

Phase I: Programming

Part A Due: _____   Part B Due: _____   Part C Due: _____

A. Research various shapes, sizes, codes, arrangements, and clearances for the above-outlined spaces. Research fixture, furniture, and equipment for the same. Research the merchandise classifications planned for the store. Analyze the program outlined above and develop a programming matrix keyed for adjacency locations. A sample blank matrix is provided in Chapter 31.

B. With the information gathered from the matrix, provide a minimum of three bubble diagrams for the spatial relationships of the areas required.

C. Provide a one-page typed concept statement of your proposed planning direction and initial design concept.

Presentation formats for Parts A through C: White bond paper, 8½″ × 11″, with a border. Insert in the binder or folder. The matrix and bubble diagrams are to be completed on the CADD program.

Evaluation Document: Programming Evaluation Form

Phase II: Conceptual Design

Part A Due: _____   Part B Due: _____

A. Develop a preliminary partition and department location plan, showing all walls, ceilings, built-in fixtures, fitting rooms, doors, and windows. Label departments and rooms and indicate the square footage proposed. These plans can be freehand drawings or CADD drawings in 1/4″ scale. Drawings must have an appropriate border. Finished drawings must be a blueprint or white bond copy.

B. Develop the above plans into a final floor plan, showing all items noted in the program. This plan is to be completed using the CADD program. Develop ceiling and lighting studies into a final reflected ceiling plan, showing all ceiling conditions and heights on the CADD program. Plans or elevations are to include stereo speakers, security cameras, sprinkler head locations, emergency lighting, and exit signs as required.

Evaluation Document: Conceptual Design Evaluation Form

Phase III: Design Development

Part A Due: _____   Part B Due: _____

A. Develop a color, material, and final fixture scheme. Samples and pictures can be presented in a loose format.

B. Develop a perspective sketch of the Sportswear Department and an elevation of the Clothing Department, showing all design elements and furnishings.

Evaluation Document: Design Development Evaluation Form

Phase IV: Final Design Presentation   Date Due: _____

A. The final presentation will be an oral presentation to a panel jury. Project must be submitted on 20″ × 30″ illustration boards. Your name, the date, the project name and number, the instructor's name, and the course title must be printed on the back of each board in the lower left-hand corner. Submit as many boards as required, including a colored rendering, colored elevation, fixture and furniture selections, material selections, and plans.

B. The plans should be presented in 1/4″ scale. The elevations are to be presented in 1/2″ scale.

Evaluation Document: Final Design Presentation Evaluation Form

Project Time Management Schedule: Schedule of Activities, Chapter 31.

The AutoCAD LT 2002 drawing file name is ultimamoda.dwg and can be found on the CD.

REFERENCES

Books

See previous retail projects.

Trade Magazines

See previous retail projects.

Fixture Manufacturers

See previous retail projects.

Special Resources

See previous retail projects.

Project References

Ermenegildo Zegna: http://www.ezegna.com
Jhane Barnes Menswear: http://www.jhanebarnes.com
Perry Ellis: http://www.perryellis.com
Salvatore J. Cesarani: http://www.cesarani.com

NOTES

CHAPTER 12

Escape to Maui + Panama Jacks

OUTCOME SUMMARY

The project features three phases of the design process: identifying and analyzing the client's needs, developing conceptual skills through schematic design concepts, and refining ideas from the schematic design phase. The project focuses on concepts for a theme-oriented mass merchandise, and resort retailer. The existing site conditions, which include floor level changes, provide accessibility and safety challenges.

client profile

Ed and Gail own Maui Clothing Company, Inc., a resort wear retailer of the mass merchandising type, that consists of ten stores located solely in the Hawaiian Islands. They have signed a ten-year lease for a space at the Lahaina Cannery Mall for a shop called Escape to Maui + Panama Jacks, and the design will be used on the next three stores. The retailer's strength is moderate-priced resort wear, so it makes perfect sense that they do

Figure 12–1 Mall Logo

not buy fall fashions or colors. The merchandise is colorful and fashionable and includes a great swimwear business. The majority of customers are 30 to 45-year-old yuppies, however; during the summer months the buyer is younger and includes the 20 to 25-year-old.

Designed by MCG Architects and Murayama Kotake Nunokawa & Associates Inc., Lahaina Cannery is Maui's only fully enclosed, air-conditioned mall. It is a one-story 124,121 square foot visitor- and tourist-oriented retail mall with 61,715 square feet of community anchors and 62,406 square feet of specialty stores and restaurants. The center is owned by the Hawaii Omori Corporation and boasts an exceptional international food court. Hawaiian artists, cultural events, and festivals are enjoyed year round by visitors and residents.

Formerly a pineapple cannery facility, it was restored and converted to a mall in 1987. The unique historic architecture was maintained—the original corrugated metal facing was moved to the interior, and skylights and clear stories were added to flood light into the interior of the structure. The mall storefront design elements are reminiscent of old Lahaina Town's Front Street and boat harbor.

The client is leasing two spaces adjacent to each other for the shops. Space #A3, called Escape to Maui, contains the women's merchandise and space #A4, called Panama Jacks, is the men's store. There is to be no physical wall separation between the shops on the interior. The shops share the same cash wrap facility, and the main aisle must pass through both spaces. This sounds simple enough; however each space has a different floor elevation. Space #A4 is 8″ lower than #A3. You have to address accessibility and safety when planning ramps and steps, and carefully plan where the transitions occur between the different groups of merchandise.

Gail wants you to carry a theme through both stores to create a tropical feeling of being on Maui. She has suggested a color theme of sand and white for the entire store (including the flooring), light pine or bleached oak materials, and palm trees. It is important that the space be very airy and light; therefore, consider incorporating luminous ceiling treatments and use ample amounts of both fluorescent and track lighting. The client is quite definite regarding fixture flexibility, type, and function, and the use of slat wall is required. The walls must be mass merchandised with face-out bars at the triple hang top height to present the merchandise. They use T-stands and four-way and round metal racks extensively, and they must match the design theme. Signs on the top of the merchandise racks will also be used.

Figure 12–2   Exterior rendering of the mall

project details
General and Architectural Information

Location: Escape to Maui + Panama Jacks, Spaces #A3 and #A4, Lahaina Cannery Mall

Gross square footage: 2,600. Net usable square footage: 2,450

Initial number of employees: Four. Projected number of employees: six in one year

Hours of operation: Monday–Sunday: 9:30 A.M. to 9:30 P.M.

Figure 12–3   Photo of storefront area from mall interior

Project budget: $180,000 for fixtures. The contractor has been selected, and the contract will be executed on a time and material basis.

Opening or move-in date: Grand opening, January 15

New construction. Mall facility plans are available. Site information gathering is required.

Consultants required: Architect, electrical engineer, and mechanical engineer

Security: Built-in Sensormatic equipment is required at the entrance. Security cameras will be used, and the mall provides 24-hour security. Lock devices are not required on fixtures, excluding the fitting rooms, showcases, and cash drawers.

Codes and Governing Regulations

Occupancy: Mercantile

Building Type: Type II. Fire Resistive

Design the project to comply with the most current edition of the ICBO Uniform Building Code (UBC), the National Electrical Code (NEC), the Uniform Fire Code (UFC), and the Americans with Disabilities Act. The Islands of Hawaii are presently rated as zone 3 seismic regions. The UBC contains information from the seismological and structural engineering communities to reduce the risk of damage from earthquakes and windstorms.

Exit signs throughout and accessible fitting rooms and cash wraps are required.

Fire Protection: Automatic Fire Sprinkler Suppression System

Facilities Required

Storefront, Cash Wrap, Employees' Lunchroom and Stockroom. Escape to Maui: Women's Sportswear, Swimwear, Aloha Shop, Accessories, and Fitting Rooms. Panama Jacks: Men's Sportswear, Swimsuits and Shorts, T-shirts, Aloha Shop, Accessories, and Fitting Rooms

Spatial Requirements

STOREFRONT AND SHOW WINDOW

- Provide two entries: one for Panama Jacks (8′ 4″ jamb opening width, 8′ 8″ opening height) and one for Escape to Maui (15′ 0″ jamb opening width, 8′ 0″ opening height). Each entry must have an overhead-coiling door. Conceal the open door, housing, and motor in a space above the door in the ceiling. Recess any steel tracks on each side of the door openings.

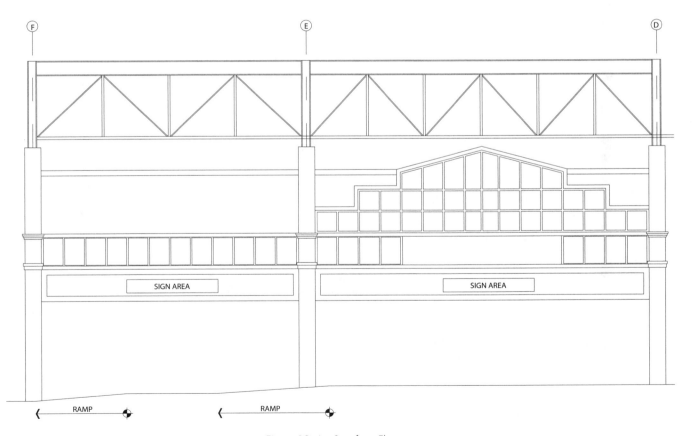

Figure 12–4   Storefront Elevation

- The balance of the Storefront must be all glass. Any mannequins used in the Show Window area should be a highly stylized type. Gail especially likes the flat cutout shape of a figure with no identifiable face features that she has seen in other stores.
- A new logo must be developed for each shop and incorporated into the sign area of the Storefront.
- Hard surface flooring is required at the entrances and on the primary store aisles. They have requested that the balance of the flooring be carpet.

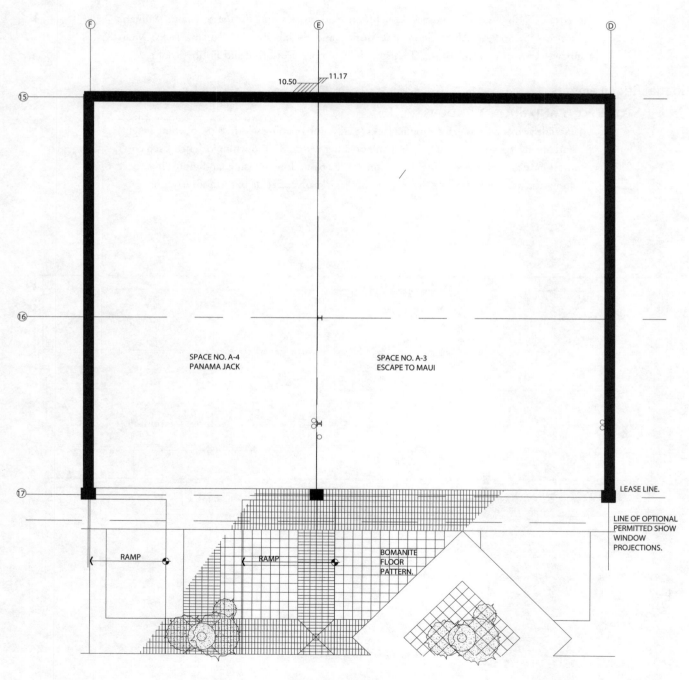

Figure 12–5 Lease Plan

PANAMA JACKS MEN'S SHOP

Planning

- 1,100 square feet. Space #A4
- Cash Wrap and Aloha Shop should be transition elements between Panama Jacks and Escape to Maui.
- All merchandise is hanging and includes:
  - Sportswear: 50% of sales floor. Adjacency: Entrance and Swimsuits and Shorts. Classifications include woven sport shirts, knit polo shirts, and pants.
  - Swimsuits and Shorts: 20% of sales floor. Adjacency: Sportswear and T-shirts
  - T-shirts: 22% of sales floor. Adjacency: Swimsuits and Aloha Shop
  - Aloha Shop: 5% of sales floor. Adjacency: T-shirts. Locate at rear of the sales floor next to the women's Aloha Shop. This department consists of Hawaiian wear and includes aloha print shirts and swimsuits for men.
  - Accessories: 3%. Adjacency: Cash Wrap. Merchandise includes hats, belts, socks, and beach games such as Smash Ball.
- Provide 3 to 4 fitting rooms a minimum size of 3′ × 4′. One fitting room is to be handicap accessible. Each fitting room must have a mirror and two garment hooks. For security reasons, fitting room doors must have locks. The fitting room doors will be locked and the sales staff will provide customer access, because a key is required for entry.
- Provide at least two mirrors on the sales floor.

ESCAPE TO MAUI WOMEN'S SHOP

Planning

- 1,350 square feet. Space #A3
- Cash Wrap and Aloha Shop are to be transition elements between Panama Jacks and Escape to Maui.
- All merchandise is hanging and includes:
  - Sportswear: 40% of sales floor. Adjacency: Entrance and Swimwear. Classifications include Bali fashions, pants, shorts and top separates, shorts and top sets, jumpsuits, rompers, and better dresses.
  - Swimwear: 25% of sales floor. Adjacency: Sportswear and Fitting Rooms. Classifications include swimsuits and swimsuit cover-ups.
  - Aloha Shop: 25% of sales floor. Adjacency: Swimwear. Locate at rear of the sales floor next to the men's Aloha Shop. This department consists of Hawaiian wear and includes long and short muumuu dresses, aloha print jumpsuits, rompers, shorts sets, and blouse and short separates for women.
  - Accessories: 10% of sales floor. Adjacency: Cash Wrap. Classifications include jewelry, handbags, beach bags, belts, sandals, slippers, hats, and beach towels. Costume jewelry must be in the showcases at the Cash Wrap for security purposes.
- Provide 5 to 6 fitting rooms a minimum size of 3′ × 4′. One fitting room must be handicap accessible. Each fitting room must have a mirror and two garment hooks. For security reasons, fitting room doors must have locks. The fitting room doors will be locked and the sales staff will provide customer access, because a key is required for entry.
- Provide at least three mirrors on the sales floor.

Equipment

- For both shops: Provide public address (P.A.) System speakers in the ceiling. Taped or radio music will be played in the store.

Design

- For both shops: All walls must be high enough to display face-out merchandise above the hanging merchandise (double height hang bars with face-out above). Approximately 12′ of the wall height is to be constructed of slatwall. The stores are stocked at maximum capacity.
- Wall finishes can be paint or medium-weight vinyl.

Lighting

- For both shops: The general lighting should be recessed fluorescent fixtures with recessed incandescent fixtures at the entrance.
- Provide continuous, 1′ × 4′, recessed fluorescent fixtures along the perimeter walls to light the merchandise and Cash Wrap desk.
- Provide track lights to highlight feature walls, above the Cash Wrap and in the front window displays.
- Locate emergency lighting and illuminated exit signs.

Cash Wrap

- ADA requires that cash wraps be on an accessible route and double cash wraps are to be a minimum of 36″ apart.
- Locate the Cash Wrap in a central place between the men and women's shops. It will service both.
- A removable carpet is required behind the Cash Wrap. Selection must provide underfoot comfort for the sales staff and be easily maintained or replaced.
- Provide several showcases to display accessories and incorporate them into the Cash Wrap fixtures.
- Electrical equipment in this area will be two computerized POS terminals, price scanners, calculator/adding machine, and two telephones.

Employees' Lunchroom

- 60 square feet. Adjacency: Stockroom. Locate at rear of shops.
- A table and four chairs are required.
- Provide an electrical junction box for future office use and a telephone.

Stockroom

- 60 square feet. Adjacency: Employees' Lunchroom. Locate at rear of shops.
- Provide fixtures for hanging merchandise and store supplies.
- Locate the electrical and telephone service panel in this space. Provide an electrical junction box for future office use and a telephone.

SUBMITTAL REQUIREMENTS

Phase I: Programming

Part A Due: _____   Part B Due: _____   Part C Due: _____

- A. Research various codes, arrangements, and clearances for the above-outlined spaces. Research fixture, furniture, and equipment for the same. Analyze the program outlined above, and develop a programming matrix keyed for adjacency locations. A sample blank matrix is provided in Chapter 31.

- B. With the information gathered from the matrix, provide a minimum of three bubble diagrams for the spatial relationships of the areas required.

C. Provide a one-page typed concept statement of your proposed planning direction and initial design concept.

Presentation formats for Parts A through C: White bond paper, 8½″ × 11″, with a border. Insert in the binder or folder. The matrix and bubble diagrams are to be completed on the CADD program.

Evaluation Document: Programming Evaluation Form

Phase II: Conceptual Design

Part A Due: _____   Part B Due: _____   Part C Due: _____

A. Provide a preliminary partition and department location plan, showing all walls, ceilings, built-in fixtures, fitting rooms, doors, and windows. Label departments and rooms and indicate the square footage required and the square footage proposed. These plans can be freehand drawings or CADD drawings in 1/4″ scale. Drawings must have an appropriate border. Finished drawings must be a blueprint or white bond copy.

B. Develop the above plan, including a preliminary ceiling, fixture, furniture, and furnishings layout of each department and room. Label departments and rooms and indicate the square footage required and the square footage proposed. This plan can be a freehand drawing or CADD drawing in 1/4″ scale. Drawings must have an appropriate border. Finished drawings must be a blueprint or white bond copy.

C. Develop the above plans into a final floor plan, showing all items noted in the program. This plan must be completed using the CADD program. Develop ceiling and lighting studies into a final reflected ceiling plan showing all ceiling conditions and heights on the CADD program. Plans or elevations are to include stereo speakers, security cameras, sprinkler head locations, and emergency lighting and exit signs as required.

Evaluation Document: Conceptual Design Evaluation Form

Phase III: Design Development   Part A Due: _____   Part B Due: _____

A. Develop a color, material, and a final fixture scheme. Samples and pictures can be presented in a loose format.

B. Develop a perspective sketch of the interior of the store, showing all design elements and furnishings. Develop a storefront elevation and section, showing all design elements.

Evaluation Document: Design Development Evaluation Form

Phase IV: Final Design Presentation   Date Due: _____

A. The final presentation will be an oral presentation to a panel jury. Project must be submitted on 20″ × 30″ illustration boards. Your name, the date, the project name and number, the instructor's name, and the course title must be printed on the back of each board in the lower left-hand corner. Submit as many boards as required, including a colored rendering, storefront elevation, fixture and furniture selections, and material selections and plans.

B. The plans should be presented in 1/4″ scale. The elevations should be presented in 1/2″ scale.

Evaluation Document: Final Design Presentation Evaluation Form

Project Time Management Schedule: Schedule of Activities, Chapter 31.

The AutoCAD LT 2002 drawing file name is maui.dwg and can be found on the CD.

REFERENCES

Books

See previous retail projects.

Trade Magazines

See previous retail projects.

Trade Organizations

See previous retail projects.

Fixture Manufacturers

See previous retail projects.

Special Resources

See previous retail projects.

Project References

The Cookson Company: http://www.cooksondoor.com
Lahaina Cannery Mall http://www.lahainacannerymall.com
The Overhead Door Company: http://www.overheaddoor.com
Porvene Doors http://www.porvenedoors.com

CHAPTER 13

Cherry Creek

OUTCOME SUMMARY

This project brings into focus what is required to work with corporate fixture guidelines and corporate design criteria and includes skill development in retail program analysis and planning and fixture development. It is a Charette that will provide experience in working within a short three-hour time limit concentrating on the planning phase of a Fine Jewelry Department within a major department store.

client profile

May D&F is a full-line department store with sixty-three locations and is owned by May Department Stores Company. The merchandise line highlights designers such as Tommy Hilfiger, Ralph Lauren, Dana Buchman, Ellen Tracy, Jones New York, DKNY, Calvin Klein, Polo, Nautica, Claiborne, and Perry Ellis. The departments included in May D&F are Jewelry; Cosmetics; Fragrance; Accessories; Men's, Women's, and Children's Apparel; Intimate Apparel; Luggage; Bedding and Linens; Housewares; China; Silver; and Furniture. You are to plan and design the Fine Jewelry Department that is situated on the first level adjacent to the escalator.

The store is a major anchor located in the Cherry Creek Shopping Center along with Neiman Marcus, Saks Fifth Avenue, and Lord & Taylor. "Cherry Creek, upon opening in August 1990, established itself as the only world-class, quality shopping experience in a vast six-state region. Featuring the region's only Neiman Marcus and Saks Fifth Avenue—along with such exclusive stores as Polo Ralph Lauren, Louis Vuitton, Mont Blanc, Burberrys, Hyde

Figure 13–1   Escalator Well Opening, Second Level. (All photographs courtesy: May D & F; Cole Martinez Curtis & Associates, Dennis Takeda, Carl Turnbull; Durant/Flickinger Associates; Robert Pisano, The Agree Corporation.)

Park and Pottery Barn—the center draws from that regional market as well as Denver's most affluent communities."[1] The Taubman Company owns the center. According to their demographic research, the community population is 1,750,000 with an average household income of $67,500 and average age of 36.

project details
General and Architectural Information

Location: May D&F, Fine Jewelry Department, First Level Escalator. Cherry Creek Shopping Center

Gross square footage: 1,200 Fine Jewelry

Hours of operation: Monday –Friday: 10 a.m. to 9 p.m.; Saturday: 10 a.m. to 8 p.m.; Sunday: 11 a.m. to 6 p.m.

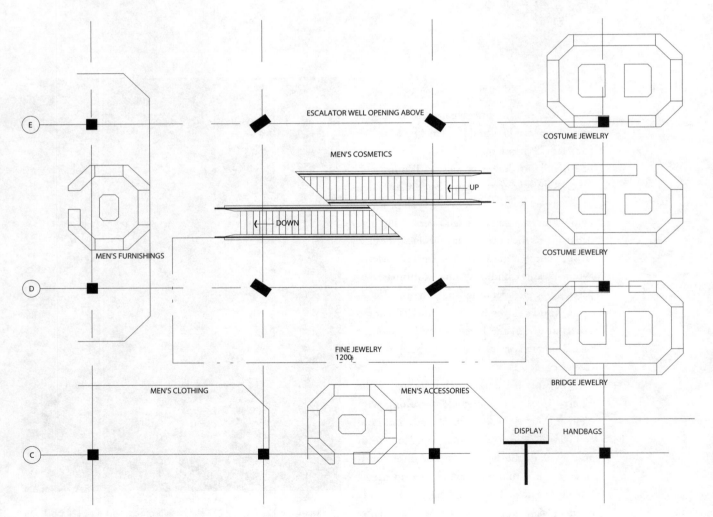

Figure 13–2   Floor Plan with Adjacent Departments

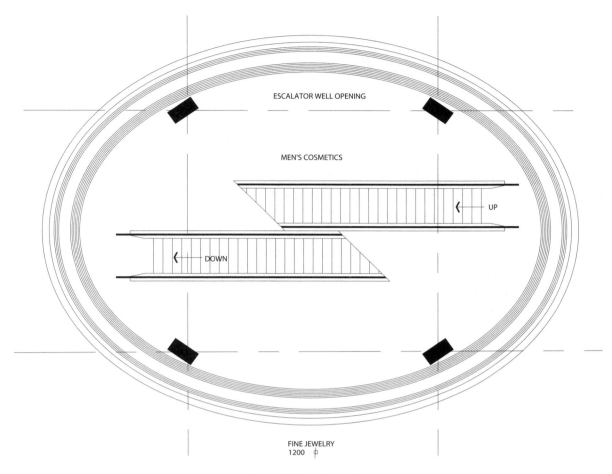

ESCALATOR WELL OPENING

MEN'S COSMETICS

← UP

← DOWN

FINE JEWELRY
1200   φ

Figure 13–3   Ceiling Plan

Facilities Required

First Level: Fine Jewelry. Adjacent departments are Bridge and Costume Jewelry, Accessories, Men's Accessories, and Men's Cosmetics. You are required to plan the Fine Jewelry Department following the guidelines below.

CORPORATE PLANNING GUIDELINES

- The escalator well architectural elements, including ceiling configurations, column details, and escalator, are indicated in Figure 13–1 and Figure 13–5. The floor plan showing the adjacent departments is indicated in Figure 13–2. Figure 13–1 illustrates the second level escalator area. Your solution must be integrated with both the plan and the architectural elements.

- The Fine Jewelry Department plan limits are indicated on Figure 13–9. Use this plan to develop and indicate your final solution.

- Provide approximately 86 linear feet of showcases, including a back case solution using the allowed floor space and escalator wall. Showcase glass is 1/3 vision (glass is 12″ high) indicated in Figure 13–4. The case is 1′ 10″ deep and it has a 4″ high base. In addition, provide two to three museum-type display cases. Indicate linear footage of each showcase island (or group) on the plan.

- Passage openings between showcases with an access to back cases are to be 3′ wide and located on secondary aisles. This opening should occur one case in from the corner showcases where possible.

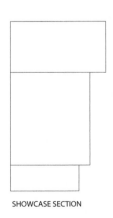

SHOWCASE SECTION

Figure 13–4
Showcase Profile

- Provide 3′ aisles between showcases and back island and showcase passage openings. The minimum floor spacing between loose floor fixtures (museum display cases in this instance) is 3′.
- Main aisles must be a minimum of 7′ 0″ and secondary aisles between islands 5′ 0″.

ADA GUIDELINES

- One showcase island must comply with ADA guidelines.
- ADA requires that complying showcase islands be on an accessible route.
- Provide 36″ between showcases and back islands, between multiple back islands, and 36″ minimum opening into showcase islands.
- Provide a removable service counter and storage for it.

CORPORATE FIXTURE GUIDELINES

- Heights of back cases are as follows:
  - Island type: 5′ 6″ maximum height
  - Against escalator: 8′ 0″ maximum height with a flat top surface
  - Escalator well opening: 6′ 0″ maximum height with a flat top surface
- Showcase lengths are typically 5′ long. You may fill in with 6′ long cases as needed.

Furniture

- One lamp is required at each corner case.
- Provide countertop mirrors for every three showcases.

Equipment

- A duplex outlet is required in every corner showcase for lamps.
- Provide power in back cases for transparencies.
- Provide power for lighting that is required in showcases and back cases.
- Indicate the power requirement locations on the plan. Provide a floor junction box for each run of showcases and back cases.

Design

- The escalator design elements and ceilings are indicated in Figure 13–1 and Figure 13–5, and your solution must be integrated with the overall visual scheme. Architectural materials include painted gypsum board wall and ceiling surfaces, custom bird's-eye maple column enclosures and second level handrails, wheat and dark perlino rosato marble floors, and glass-panel escalator balustrades.
- The corporate Visual Department will select the case pad fabric.
- A lighting consultant will provide the lighting plan.

Figure 13–5
Photograph of ceiling at the escalator above the Fine Jewelry Department

Figure 13–6
Shoe Department

Figure 13–7
Fur Department

Figure 13–8
Better Sportswear Department

SUBMITTAL REQUIREMENTS

Phase I: Programming

A. Read the background information, client's program requirements, and limitations, and analyze the problem.

B. Formulate some visual ideas to express the character of the space. Use the scaled floor plan in Figure 13–9 and sketch your ideas on tissue overlays.

Evaluation Document: Programming Evaluation Form

Phase II: Conceptual Design

A. Develop a few simple tissue sketches of the fixture plan before finalizing a floor plan.

B. Prepare a final floor plan on the plan provided in Figure 13–9. This plan may be drawn freehand or using drafting equipment but it must be drawn to 1/8″ scale.

C. Draw a freehand perspective sketch of the most important view of the department on tissue paper.

D. At the end of the three-hour time limit, present your solution to the class. Your verbal presentation must include justification for your spatial and functional solutions, as well as the visual character of the space designed.

Evaluation Document: Conceptual Design Evaluation Form

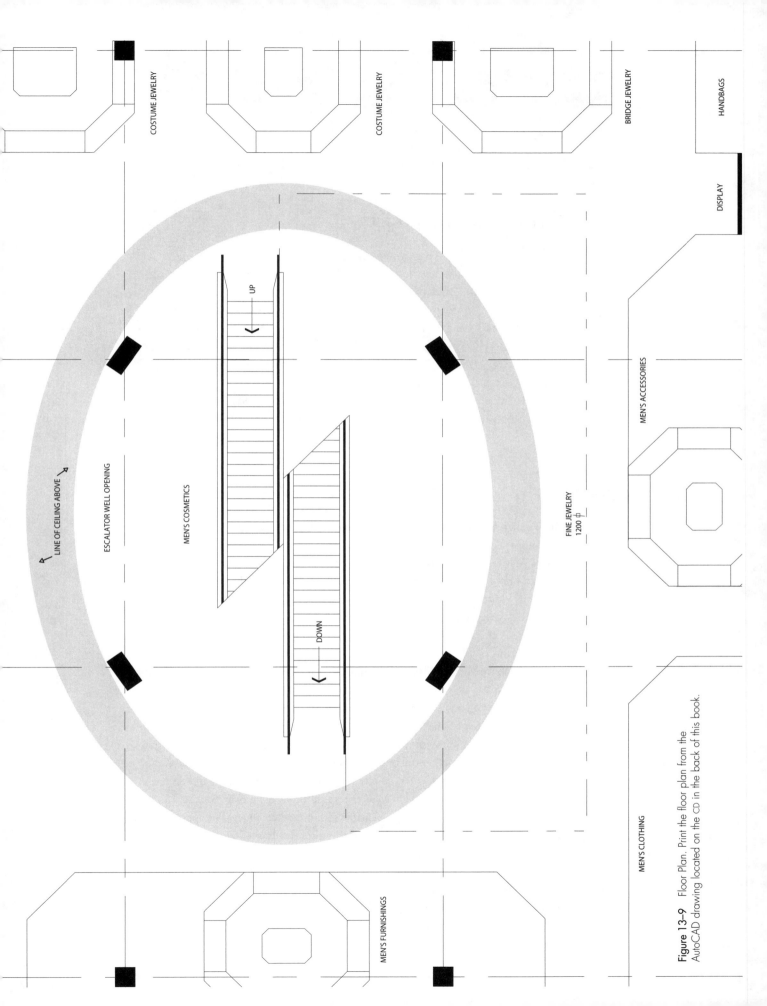

COSTUME JEWELRY

COSTUME JEWELRY

BRIDGE JEWELRY

HANDBAGS

DISPLAY

UP

LINE OF CEILING ABOVE

ESCALATOR WELL OPENING

MEN'S COSMETICS

MEN'S ACCESSORIES

FINE JEWELRY
1200

DOWN

MEN'S CLOTHING

MEN'S FURNISHINGS

Figure 13–9 Floor Plan. Print the floor plan from the AutoCAD drawing located on the CD in the back of this book.

CHAPTER 14

Christopher Joseph

OUTCOME SUMMARY

The project features a team approach to three phases of the design process: to define team responsibilities; to discuss goals and explore client requirements through design team meetings, buzz sessions, and dialogue groups; and to establish optimal space plans using adjacency diagrams, which show the proximity of the various merchandise groups. Refinement of design concepts must be conveyed through drawings, sketches, material samples, and space plans for a final design presentation. The team is responsible for researching the local codes and arranging a preliminary plan check meeting with local code officials to review life safety and accessibility.

client profile

A California retailer, Christopher Joseph, has premier locations in Beverly Hills, Costa Mesa, San Diego, San Francisco, Palo Alto, and Las Vegas, some created by design icons such as Skidmore, Owings, & Merrill. Their plans are to open another location in the heart of Los Angeles. CJ is primarily a fine women's specialty store filled with inspirational fashion modes. It offers a variety of designer clothing and features quite a few bright new design talents. It includes only five retail categories—women's apparel, cosmetics, fragrances, accessories, and a small men's apparel department. Its customers have an eye for value and a high regard for the individual service that is characteristic of CJ, which includes a personal shopper. The customer profile includes consumers with an above-average disposable income, the professional woman, and tourists alike because of its location.

The new store is located on the west side of Los Angeles in the Century City Shopping Center, a Westfield Shoppingtown. The Master Plan was developed by Welton Becket & Associates; other architects responsible for its vision are Johnson, Faine, and Periera Associates and Minoru Yamasaki. Century City is a chic, central business district that also includes five star accommodations, the ABC Entertainment Center, quality fashion retailers, nightclubs, and dining in a distinctive setting. Its name alone says a lot. In fact, it was formerly the movie "back lot" for 20th Century Fox film studios and you will spot celebrities shopping there today. The approach from Santa Monica or Olympic Boulevard gives a first time visitor the dramatic impression that you are driving right into a movie set because of the stark contrast of the low urban landscape of the surrounding community and the adjacent mini-metropolis. The setting has all the hallmarks of a 1958 Futuristic City, with streets named Galaxy Way and Constellation Boulevard, fountains, soaring office towers, and pedestrian bridges. The pedestrian bridge over the Avenue of the Stars joins the shopping center with the modern office towers.

Most of the center is outside and includes 140 one-of-a-kind retailers, twenty-two dining opportunities, and the famed Festival Marketplace. Tiffany & Co., Abercrombie & Fitch, Banana Republic, MaxMara, Ann Taylor, Pottery Barn, Restoration Hardware, BOSS Hugo Boss, and J. Crew are a few of the fine retailers in the mall. Bloomingdale's West Coast flagship store and Macy's are the major anchors, and Gelson's Gourmet Market is an exceptional highlight of the mall.

The exterior of the storefront will be renovated. The entrance façade has been redesigned to incorporate an all-glass entrance system including a skylight that creates an expansive feeling at the entry. There is a clearstory skylight in the center of the space serving to capture the incoming daylight. You are to make optimal use of its 29,000 square feet. The design should be out of the ordinary and the merchandise presented in a straightforward and simple manner. Make a style statement that is young, lively, and experimental. In short, have fun with the space; incorporate unusual, witty, and dramatic design elements.

CJ is looking for an exceptionally unique merchandise presentation solution. Custom fixtures are to be designed for the store to distinguish CJ from its competitors. Keep displays to a minimum so that the merchandise stands out. To give it a luxurious feeling, features are to include plenty of sitting spaces, and the fitting rooms should be spare but elegant.

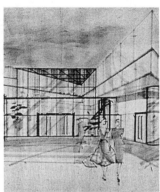

Figure 14–1
New Storefront Preliminary Sketch

In general, establish floor, wall, and ceiling configurations, including prominent design features. Pay particular attention to sight lines and department depths. Department depths of over 40 feet are undesirable. The circulation scheme should be continuous and create a progression of exciting viewpoints as the customer is lead through the space. Design all perimeter fixtures and floor fixtures. Manufactured fixtures may be included. Select all materials and finishes and be certain that life safety and fire codes are met. Commercial grade products are required for safety, maintenance, and durability. Also develop a signage system for directing customers to support services and non-retail facilities.

Following is a list of client guidelines that should be adhered to in developing the project. Aisles at the entrance are to be a minimum of 7′ 0″, and secondary ailes, 5′ 0″. Provide a minimum clearance of 3′ 0″ between fixtures and allow for the space that the garments will occupy as well. Do not use round metal fixture racks. Cash wraps should be located to visually cover the sales floor and service the fitting rooms but not placed in front of feature design elements. You are required to locate and design the cash wraps but are not required to fully develop the interior functions of the wrap. This is applicable to all of the cash wraps in the store. And remember that mirrors are to be advantageously located throughout the store.

On the plan you must indicate department names, merchandise classifications, major vendors, requested square footage, proposed square footage, and linear footage of each showcase group.

project details

General and Architectural Information

Location: Christopher Joseph, Century City Shopping Center, 10250 Santa Monica Boulevard, Century City, CA 90067

Gross square footage: 29,000 First Level; 2,000 Mezzanine. 31,000 Total

Net usable square footage: 20,600 First Level; 1,400 Second Level. 22,000 Total

Hours of operation: Monday –Friday: 10 a.m. to 9 p.m.; Saturday: 10 a.m. to 9 p.m.; Sunday: 11 a.m. to 6 p.m.

Consultants required: Architect, lighting consultant, electrical engineer, and mechanical engineer

Security

- Built-in Sensormatic equipment is required at the entrance. Locks are required on all showcases.
- Closed-circuit television and electronic sensor devices are monitored in the Security Office.
- Ten observation towers are required. Locate to provide maximum coverage of the sales floor and cash wraps. The tower consists of a space of approximately 6 square feet concealed from, but with vision of, the sales floor. A ladder is required because the height of the observation window (two-way mirror) varies and should be located close to the ceiling level.

Codes and Governing Regulations

Occupancy: M Mercantile

Building Type: Type I (protected) noncombustible assembly

Fire Protection: Fire Sprinkler Suppression System

Design the project to comply with the most current edition of:

- City of Los Angeles Building Code, which is based on the California Building Code (CBC) and the ICBO Uniform Building Code (UBC) with LA City Amendments.
- NFPA National Electrical Code (NEC), City of Los Angeles Electrical Code, which is based on the California Electrical Code (CEC) with LA City Amendments.
- Title 19, fire department approval is required. Special seismic life safety issues apply to construction because the structure is in a risk level 4 seismic zone. The UBC contains information from the seismological and structural engineering communities to reduce the risk of damage from earthquakes.
- Americans with Disabilities Act, which requires that complying showcase groups and cash wraps be on an accessible route and double cash wraps are to be 36″ apart. Provide 36″ between showcases and back cases, between multiple back cases, and 36″ minimum opening into showcase groups.

    Exit signs throughout and accessible fitting rooms and cash wraps are required.

FACILITIES REQUIRED

First Level: Men's Shop, Cosmetics, Jewelry, Accessories, Handbags, Women's Shoes, Charles Jourdan Shop, Trend, Sportswear, Designer Sportswear, Dresses, Coats, and Intimate Apparel. The building function and nonselling areas include: Show Windows, Elevators, Stairs, Restrooms, Gift Wrap, Display Workroom, Security Office, and Receiving.

Second Level: (This project excludes the planning and design of the second level.) General Office, Receptionist, Office Manager, Store Manager, Assistant Manager, Operation Manager, Employee Lounge, Training Room, POS Equipment Room, and Alterations

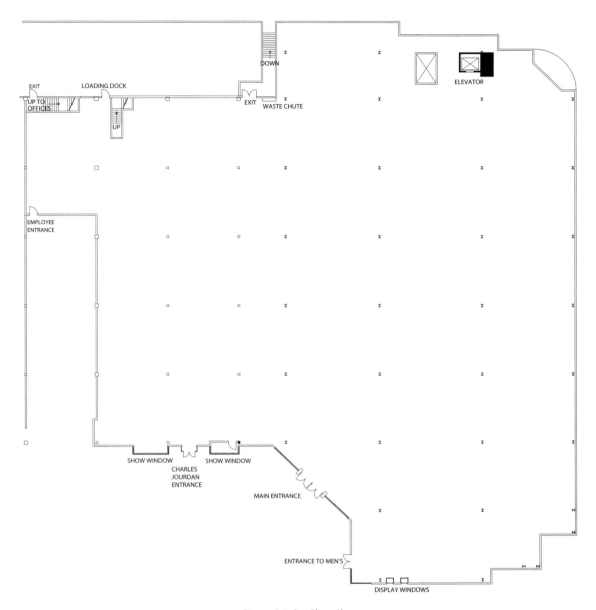

Figure 14–2 Floor Plan

First Level Spatial Requirements

MEN'S SHOP

Planning

- 1,875 square feet. Primary adjacency: Entrance. Secondary adjacency: Exclusive exterior door for entry to Men's Shop
  - Sportswear: 605 square feet. Locate adjacent to Designer Sportswear. Merchandise separates include shirts, knits, sweaters, pants, sport coats, and active wear. Designers include Donna Karan, Hugo Boss, and Zanone.
  - Designer Sportswear: 220 square feet. This department should have a prominent location in the Men's Shop. Designers include Comme des Garcons Homme Plus, Costume National, Dolce & Gabanna, and Prada.

- Men's Furnishings and Accessories: 440 square feet. Adjacency: Men's Sportswear. Dress furnishings include dress shirts and neckwear. Personal furnishings include socks, underwear, sleepwear and robes. Accessories include belts, hats, gloves, umbrellas, sunglasses, small leather, and gifts. A full-vision (glass is 26″ high) showcase is required in Accessories.
- Clothing and Outerwear: 263 square feet. Merchandise includes suits, suit separates, pants, topcoats, jackets, and rainwear. Designers include Giorgio Armani, Mani, Vestimenta, and Zegna Soft.
- Stockroom: 115 square feet. This space is used for back stock of merchandise. Provide double hang bars and shelving.

- Provide one three-way mirror on the perimeter wall in Outerwear.
- Provide two single cash wraps. One cash wrap must comply with ADA requirements. The cash wraps are used to write up sales and wrap the merchandise. Storage is required for bags, tissue, and plastic garment bags.
- Provide 3 Fitting Rooms (Minimum size: 3′ 6″ × 3′ 6″). If possible, locate the fitting rooms adjacent to Sportswear and Clothing. One fitting room must be handicap accessible. Provide a mirror, comfortable chair, shelf, and garment hooks in each fitting room. The fitting rooms are to have a shared three-way mirror area that includes a fitting platform and writing ledge or surface.

Equipment

- Electrical equipment for the cash wraps in this area include a computerized cash terminal, price scanner, calculator/adding machine, and telephone.
- Power is required for showcase lights.

COSMETICS

Planning

- 1,500 square feet. Primary adjacency: Entrance. Secondary adjacency: Sportswear, Accessories, and Jewelry. This merchandise is displayed in showcases and open-sell fixtures. Some vendors require traditional showcase groupings, and others use a more current open-sell approach. Research the vendor (designer) requirements and plan accordingly.
- Remote Stockroom: 250 square feet. This space is used for back stock of merchandise. Provide shelving.
- Provide four showcase groupings or the equivalent fixture function with a cash wrap in each. A total of 210 linear feet of cases is required with full vision (glass is 26″ high). Indicate on the plan the linear footage of each showcase group. Passage openings between showcases must be 36″ wide and located on secondary aisles within the department (not on main store aisle) if possible. One showcase group must comply with ADA requirements. One cash wrap must comply with ADA requirements.
- Provide two counter stools at each showcase grouping. Countertop mirrors and lamps are required for every four showcases and at showcases with counter stools.
- Include promotional floor fixtures.
- Cosmetics: Body & Soul, Chanel, Clarins, Make Up For Ever, Nars, Philosophy, Poppy, Prescriptives, Shiseido, Shu Uemura, Stila, T. Le Clerc, and Vincent Longo. 170 linear feet of showcase is required.
- Fragrances: Annick Goutal, Bulgari, Chanel, Comme des Garcons, Dolce & Gabbana, Guerlain, Hermes, Issey Miyake, Jean Patou, Jill Sander, Romeo Gigli, and Yohji Yamamoto. 25 linear feet of showcase and 25 linear feet of open-sell showcase are required.

- Spa and Accessories: This is open-sell merchandise including spa sets, bath and body treatments, aromatherapies, and mirrors. 15 linear feet of open-sell showcase or fixtures is required.

Equipment

- Electrical equipment for cash wraps in this area include a computerized cash terminal, price scanner, calculator/adding machine, and telephone.
- Power is required for showcase lights and case-top lamps.

JEWELRY

Planning

- 400 square feet. Primary adjacency: Accessories. Secondary adjacency: Cosmetics, Handbags.
- Provide one showcase grouping (or group) or the equivalent fixture function with a cash wrap in the back case. A total of 52 linear feet of cases is required. Provide 12 linear feet for Fine Jewelry with third vision (glass is 12″ high) showcase. All other cases are half vision (glass is 18″ high). Indicate on the plan the linear footage of each showcase group. One showcase group must comply with ADA requirements. One cash wrap must comply with ADA requirements.
- Countertop mirrors and lamps are required.
- Merchandise includes fine jewelry, fashion watches, and fashion (costume) jewelry.
- Designers include Beamon, Beth Orduna, Cynthia Wolff, Dana Kellin, Debbie Fisher, Jeanine Payer, Kazuko, Maria Rudman, Mee & Ro, Mizuki, Mallory Marks, Reinstein Ross, Renee Lewis, and Timothy Watkins.

Equipment

- Electrical equipment for cash wraps in this area include a computerized cash terminal, price scanner, calculator/adding machine, and telephone.
- Power is required for showcase lights and case-top lamps.

ACCESSORIES

Planning

- 1,000 square feet. Primary adjacency: Jewelry. Secondary adjacency: Handbags. This merchandise is displayed in showcases and open-sell fixtures. A large portion of this department contains seasonal merchandise (scarves, knits, hats, gloves, sunglasses).
- Remote Stockroom: 100 square feet. This space is used for back stock of merchandise. Provide double hang bars and shelving. Provide a department manager's workstation, 24″ × 36″ surface with file storage, overhead storage, task chair, computer, and telephone.
- Provide one showcase grouping (or group) or the equivalent fixture function with a cash wrap in the back case. The showcase group and the cash wrap must comply with ADA requirements if not adjacent to Cosmetics and Jewelry. A total of 48 linear feet of cases with full vision (glass is 26″ high) is required.
- Countertop mirrors and lamps are required.
- Merchandise includes hair ornaments, sunglasses, small leather goods, belts, millinery, gloves, knits, scarves, shawls, umbrellas, rainwear, slippers, socks, hosiery (190 square feet), and bodywear.
- Designers include Anya Hindmarch, Fendi, Henry Beguelin, Henry Cuir, J & M Davidson, Julia Hill, Le Cachemerien, Lucy Barlow, Maria La Rosa, Marni, Megan Park, Prada, and Roberta di Camerino.

Equipment

- Electrical equipment for cash wraps in this area include a computerized cash terminal, price scanner, calculator/adding machine, and telephone.
- Power is required for showcase lights and case-top lamps.

HANDBAGS

Planning

- 750 square feet. Primary adjacency: Accessories. Secondary adjacency: Shoes. This merchandise is displayed in showcases and open-sell fixtures.
- Remote Stockroom: 100 square feet. Shelving is required.
- Provide one showcase grouping (or group) with a cash wrap in the back case or the equivalent fixture function. A total of 48 linear feet of full vision (glass is 26″ high) cases is required. The cases will contain evening bags and high-end handbags.
- Designers include Burberry, Isabella Fiore, Giuseppe Zanotti, Kate Spade, Marc Jacobs, Michael Kors, Prada, Ferragamo, Mabiani, Furla, Judith Leiber, and Franchi.

Equipment

- Electrical equipment for cash wraps in this area include a computerized cash terminal, price scanner, calculator/adding machine, and telephone.
- Power is required for showcase lights.

WOMEN'S SHOES

Planning

- 1,024 square feet. Primary adjacency: stockroom and Charles Jourdan Shop. Secondary adjacency: Handbags
- Women's Shoe Stock: 930 square feet. Provide a 42″ passage opening into the stock area from the sales floor, and baffle the visibility into stock from the sales floor. The stockroom should be contiguous to Jourdan stock and directly adjacent to the Women's Shoes sales area. Provide a department manager's workstation (24″ × 36″ surface with file storage, overhead storage, task chair, computer, and telephone) and a workbench in the stockroom area. A staging area of 50 square feet is also required. Provide stock shelving. Calculate the shoe pairage total provided in the stockroom and indicate on the plan. Use the following method:

    Shelving at 3′ wide × 10′ high: 46 pair per foot or 138 pair per section.

    Shelving at 4′ wide × 10′ high: 48 pair per foot or 192 pair per section.
- The required seating capacity is 30. Indicate the seat quantity on the plan. Allow 6′ aisle in front of seating groups, 5′ between seating group and fixtures, and 4′ between fixtures.
- Provide 6 fitting stools and full-length mirrors.
- One double cash wrap is required and should be located in the stockroom near the entrance. One cash wrap must comply with ADA requirements.
- Provide a wide passage opening (minimum of 8′0″ wide) into the Charles Jourdan Boutique.
- Designers include Ann Demeulemeester, Christian Louboutin, Collection Privee, Costume National, Espace, Free Lance, Gianni Barbato, Henry Cuir, Jill Sander, Manolo Blahnik, Michel Perry, Michelle Mason, Miu Miu, Prada, Prada Sport, Robert Clergerie, and Royal Elastics.

Equipment

- Electrical equipment for cash wraps in this area include a computerized cash terminal, price scanner, calculator/adding machine, and telephone. Provide two computerized cash terminals at double cash wraps.

CHARLES JOURDAN SHOP

Planning

- 955 square feet. Primary adjacency: Exterior entry doors and stock. Secondary adjacency: Main entrance and Women's Shoes
- Jourdan Shoe Stock: 340 square feet. The stockroom should be contiguous to Women's Shoes stock and directly adjacent to the Jourdan Shoes sales area. Access between Jourdan Shoe stockroom and Women's stockroom is not allowed. Use fixture shelving rather than a wall to accomplish this. Provide a 42″ opening into the stock area from the sales floor, and baffle the visibility into stock from the sales floor. A manager's office of 55 square feet is required in the stock area. Provide stock shelving. Using the method described for Women's Shoes stock, calculate the shoe pairage total provided in the stockroom, and indicate on the plan.
- You are to provide three entrances to this boutique from the following locations: exterior, main aisle of CJ Store, and Women's Shoes.
- This is a leased space and does not have to be designed beyond providing for the requirements listed above.

TREND

Planning

- 550 square feet. Primary adjacency: Main entrance
- Provide two fitting rooms. One fitting room must be handicap accessible. The fitting rooms are to have a shared three-way mirror area. Provide a mirror, comfortable chair, shelf, and garment hooks in each fitting room.
- Designers include Anna Sui, Fake London, Helmut Lang, Mui Mui, and Vivienne Tam.

SPORTSWEAR

Planning

- 9,000 square feet. Primary adjacency: Main entrance. Secondary adjacency: Trend
  - Active Sportswear: 900 square feet. Designers include Max Mara, DKNY, and Prada Sport.
  - Weekend Sportswear: 1,900 square feet. Designers include Alberta di Ferretti, Billy Blues, Bonds, Claudie Pierlot, Ghost, Juicy Couture, Petit Bateau, and Urchin.
  - Career Sportswear: 2,000 square feet. Designers include Akira, Clements Ribeiro, Fendi, Isabel Toledo, Pamela Dennis, Paul Smith, and Richard Tyler.
  - Classic Career Sportswear: 1,500 square feet. Designers include Burberry, Donna Karan, Geoffrey Beene, and Jill Sander.
  - Occasion Sportswear: 1,600 square feet. Designers include Ralph Lauren, Georgio Armani, Issey Miyaki, Costume National, Comme des Garcons, Prada, and Moschino Cheap & Chic.
  - Stockroom: 200 square feet. This space is used for back stock of merchandise. Provide double hang bars and shelving. Include two department manager workstations, 24″236″ surface with file storage, overhead storage, task chair, computer, and telephone for each.
- A total of 16 fitting rooms is required for the Sportswear Department and must be advantageously located throughout the classifications, with a minimum grouping of 4 fitting rooms per complex. Three fitting rooms are to be handicap accessible. Provide a mirror, comfortable chair, shelf, and garment hooks in each fitting room. Each complex of fitting rooms is to have a shared three-way mirror.

- Provide three single cash wraps strategically located throughout Sportswear. One cash wrap must comply with ADA requirements.

Equipment

- Electrical equipment for cash wraps in this area include a computerized cash terminal, price scanner, calculator/adding machine, and telephone.

DESIGNER SPORTSWEAR

Planning

- 1,800 square feet. Primary adjacency: Occasion Sportswear. Secondary adjacency: Dresses
- Personal Shopper: 125 square feet. This room should be located near Designer Sportswear and have access to the Designer Sportswear stockroom for customer holds. Provide a consultation room that includes a desk with file storage, desk chair, guest chair, hang bars, and a private fitting room. A stock area (5 linear feet of hanging) is needed for customer holds. Provide a three-way mirror, comfortable chair, shelf, hang bar, and garment hooks in the private fitting room.
- Stockroom: 70 square feet. This space is used for back stock of merchandise. Provide double hang bars and shelving.
- Provide 5 fitting rooms at least 6′ × 8′ in size. One fitting room must be handicap accessible. Provide a three-way mirror, fitting area with platform, sofa and chair grouping, face-out hang bars, and garment hooks in each fitting room. A shared, three-way mirror fitting area of 20 square feet is also required in the fitting room complex. It should include a fitting platform and a writing ledge or surface.
- A double cash wrap desk is required but must be located off the sales floor in the fitting room complex. One cash wrap must comply with ADA requirements.
- Include a luxurious seating arrangement.
- Designers include YSL, Dolce & Gabbana, Donna Karan, Marc Jacobs, Michael Kors, Missoni, Moschino Couture, Issey Miyake, and TSE.

Equipment

- Electrical equipment for cash wraps in this area include a computerized cash terminal, price scanner, calculator/adding machine, and telephone. Provide two computerized cash terminals at double cash wraps.

DRESSES

Planning

- 2,300 square feet. Primary adjacency: Coats. Secondary adjacency: Designer Sportswear
- Merchandise includes career dresses, weekend dresses, and occasion dresses.
- A single cash wrap is required and must comply with ADA requirements.
- Provide 6 fitting rooms (minimum size of 3′ 6″ × 3′ 6″). One fitting room is to be handicap accessible. Provide a mirror, comfortable chair, shelf, and garment hooks in each fitting room. The fitting rooms are to have a shared three-way mirror area.
- Designers include ABS, Anopia, and Carmen Marc Valvo.

Equipment

- Electrical equipment for cash wraps in this area include a computerized cash terminal, price scanner, calculator/adding machine, and telephone.

COATS

Planning

- 800 square feet. Primary adjacency: Dresses. Secondary adjacency: Intimate Apparel

- This department contains seasonal merchandise. It will swing from outerwear to incorporate more swimwear and resort merchandise in the spring season. It is important that it be located near Dresses for easy access to the fitting rooms.
- Provide one three-way mirror on the perimeter wall.
- Provide a locking fur case with sliding glass doors on the perimeter wall. Five linear feet is adequate.
- Designers include Plenty, Ramosport, Max Mara, Bill Blass, Burberry, Dana Buchman, and Marc Jacobs.

INTIMATE APPAREL

Planning

- 1,400 square feet. Primary adjacency: Coats. Secondary adjacency: Dresses.
  - Day Wear: 350 square feet
  - Sleepwear, Loungewear, and Robes: 580 square feet
  - Foundations, Bras, and Panties: 300 square feet
  - Stockroom: 60 square feet. This space is used for back stock of merchandise. Provide double hang bars and shelving.
- Robes are to be located at the back of the department.
- Provide a single cash wrap that complies with ADA requirements.
- Provide 4 fitting rooms. Locate fitting rooms adjacent to Foundations, Bras, and Panties. One fitting room must be handicap accessible. Provide a mirror, comfortable chair, shelf, and garment hooks in each fitting room. The fitting rooms are to have a shared three-way mirror area.
- Designers include Chantelle, Donna Karan, Escada, La Perla, Hanro, Le Jaby, Bella Sera, Flora, and Zen.

Equipment

- Electrical equipment for cash wrap in this area include a computerized cash terminal, price scanner, calculator/adding machine, and telephone.

RESTROOMS

- 259 square feet. Adjacency: Gift Wrap and Elevators preferred but not required. Attendant will be on duty in the restrooms.
- Doors and baffles should be located so that no one can see directly into the restroom.
- Check codes to determine the number of water closets, urinals, and lavatories required. A family-type restroom with a lounge area must be planned. Comply with ADA requirements.
- Provide full-height wall partitions and doors in each water closet location.
- Provide mirrors over lavatories with proper lighting.
- A full-length mirror is required.
- When choosing materials, consideration must be given to ease of maintenance and resistance to damage and moisture.
- Provide baby changing stations.
- Provide a lounge area and separate makeup counter with proper lighting.

GIFT WRAP

- 240 square feet. Adjacency: Restrooms and Elevators preferred but not required
- The counter, back counter, and cash wrap must comply with ADA requirements. Provide a manager workstation, 24″ × 36″ surface with file storage, overhead storage, task chair, computer, and telephone. A gift-wrap wall display panel is required.

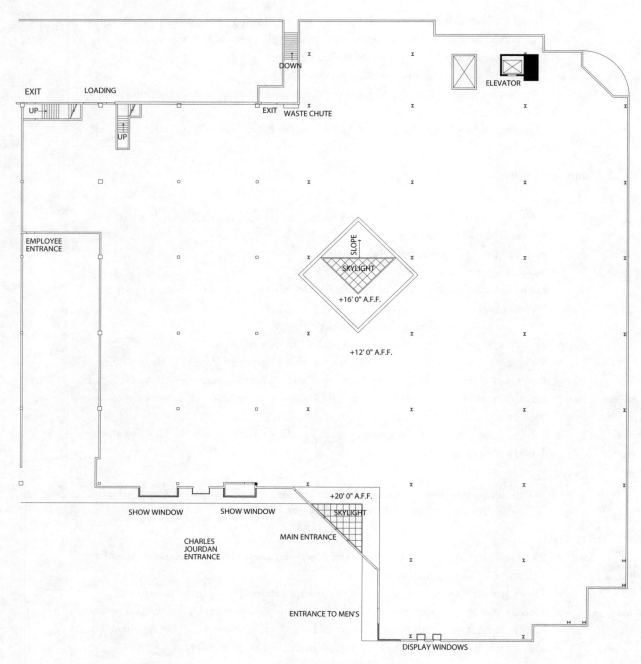

Figure 14–3   Ceiling Plan

- Provide a waiting area, including a furniture arrangement to seat four people.
- One public telephone station and a water fountain are required. Comply with ADA requirements.

DISPLAY WORKROOM

- 220 square feet. Adjacency: Receiving
- A double passage door is needed for the entrance to this space (5′ 0″ clear width opening).
- Provide a workbench; 12 linear feet of shelving; and a workstation, 24″ × 36″ surface with file storage, overhead storage, task chair, computer, and telephone.

SECURITY OFFICE

- 150 square feet. Primary adjacency: Receiving
- Plan an office with one workstation and storage cabinets. The workstation must include a work surface with return, file storage, and a desk chair.
- Incorporate closed-circuit television monitors.
- Plan an office/interview/detaining room with two private workstations. Each station must include a work surface, file storage, bookshelf, chair, and side chair.

RECEIVING

- 280 square feet. Primary adjacency: Exterior Loading Dock. Secondary adjacency: Security
- This space is used for receiving merchandise. If space allows, provide a small work counter and a stool. A wall-mounted telephone is required.
- Stockroom: 80 square feet. This space is used for holding merchandise. Provide a double hang bar and shelving. If space allows, provide a small work counter and chair.

SUBMITTAL REQUIREMENTS

Phase I: Programming   Part A Due: _____   Part B Due: _____

Part C Due: _____   Part D Due: _____   Part E Due: _____

Part F Due: _____

A. Your instructor will organize the design team (space planner, merchandiser, designer, and color and material designer) and negotiate a compensation method. Verify each team member's ability to meet the time schedule for the project.

B. Research various shapes, sizes, codes, arrangements, and clearances for the above-outlined spaces. Research fixture, furniture, and equipment for the same. Research the merchandise classifications, vendors, and designer lines planned for the store.

C. Hold two design team meetings, including a buzz session and dialogue group for the project. In the buzz session, discuss the topic of space planning and propose alternative strategies. In the dialogue group session, share your knowledge from the research you have completed and arrive at a design approach for the project.

D. Analyze the program outlined above and develop a programming matrix keyed for adjacency locations. A sample blank matrix is provided in Chapter 31.

E. With the information gathered from the matrix, provide a minimum of three bubble diagrams for the spatial relationships of the areas required. Continue to develop the plan into a block plan, indicating space allocations.

F. Provide a one-page typed concept statement of your proposed planning strategy and initial design concept approach.

Presentation formats for Parts A through F: White bond paper, 8½″ × 11″, with a border. Insert in a binder or folder. The matrix and bubble diagrams are to be completed on the CADD program.

Evaluation Document: Programming Evaluation Form

Phase II: Conceptual Design   Part A Due: _____   Part B Due: _____
Part C Due: _____   Part D Due: _____

A. Provide a preliminary partition and department location plan, showing all walls, ceilings, built-in fixtures, fitting rooms, doors, and windows. This plan must show relationships, adjacencies, department areas, and the circulation aisles. Label departments and rooms and indicate the square footage required and the square footage planned. This plan can be a freehand drawing or CADD drawing in 1/8" scale. Drawing must have an appropriate border. Finished drawing must be a blueprint or white bond copy.

B. Develop the above plan, including a preliminary ceiling, fixture, furniture, and furnishings layout of each department and room. Label departments and rooms and indicate the square footage required and the square footage planned. This plan can be a freehand drawing or CADD drawing in 1/8" scale. Drawings must have an appropriate border. Finished drawings must be a blueprint or white bond copy.

C. Develop the above into a final floor plan, showing all items noted in the program. This plan is to be completed using the CADD program. Develop ceiling and lighting studies into a final reflected ceiling plan, showing all ceiling conditions and heights on the CADD program.

D. Discuss the following mock preliminary plan; check with your instructor before proceeding. Assign a plan check manager from your group who will arrange a meeting with your local authorities for a preliminary plan checking procedure as it applies to life safety and accessibility requirements for your project. As a group, have the above floor plans reviewed for code compliance. In the event your local authorities are not able to accommodate a school group request such as this, ask a code official or architect to visit your class and share his or her knowledge in reviewing your plan.

Evaluation Document: Conceptual Design Evaluation Form

Phase III: Design Development   Part A Due: _____   Part B Due: _____

A. Develop a color, material, and a final fixture scheme. Samples and pictures can be presented in a loose format.

B. Develop perspective sketches and/or elevations of the departments, showing all design elements, fixtures, and furnishings.

Evaluation Document: Design Development Evaluation Form

Phase IV: Final Design Presentation   Date Due: _____

A. The final presentation will be an oral presentation to a panel jury. Project must be submitted on 20" × 30" illustration boards, excluding the fixture plan and reflected ceiling plan. Your name, the date, the project name and number, the instructor's name, and the course title must be printed on the back of each board in the lower left-hand corner. Submit as many boards as required, including a colored elevation or rendering and fixture, furniture, and material selections of the following departments (as assigned by instructor):

1. _____

2. _____

3. _____

4. _____

5. _____

6. _____

B. The plans should be presented in 1/8″ scale. The elevations should be presented in 1/2″ scale.

Evaluation Document: Final Design Presentation Evaluation Form

Project Time Management Schedule: Schedule of Activities, Chapter 31.

The AutoCAD LT 2002 drawing file name is cj.dwg and can be found on the CD.

REFERENCES

Books

See previous retail projects.

Trade Magazines

See previous retail projects.

Trade Organizations

See previous retail projects.

Fixture Manufacturers

See previous retail projects.

Special Resources

See previous retail projects.

Project References

Annick Goutal: http://www.annickgoutal.com
Giorgio Armani: http://www.giorgioarmani.com
Burberry: http://www.burberry.com
Chanel: http://www.chanel.com
Charles Jourdan: http://www.charlesjourdan.com
City of Los Angeles, Department of Building and Safety: http://www.lacity.org/LADBS/
Clarins: http://www.clarins.com.
Costume National: : http://www.costumenational.com
Cynthia Wolff: http://www.cynthiawolff.com
Dolce & Gabbana: http://www.dolcegabanna.it
Donna Karan: http://www.donnakaran.com
Ferragamo: http://www.ferragamo.com
Furla: http://www.furla.com
Guerlain: http://www.guerlain.com
Helmut Lang: http://www.helmutlang.com
Hermes: http://www.hermes.com
Hugo Boss: http://www.hugoboss.com

Issey Miyake: http://www.isseymiyake.com
Jeanine Payer: http://www.jeaninepayer.com
Juicy Couture: http://www.juicycouture.com
Kate Spade: http://www.katespade.com
Mabiani: http://www.andreamabiani.com
Marc Jacobs: http://www.marcjacobs.com
Michael Kors: http://www.michaelkors.com
Missoni: http://www.missoni.com
Miu Miu: http://www.miumiu.com
Moschino Couture: http://www.moschino.it
Philosophy: http://www.philosophy.com
Philosophy: http://www.philosophy.it
Prada: http://www.prada.com
Ralph Lauren: http://www.ralphlauren.com
Roberta di Camerino: http://www.robertadicamerino.com
Sephora: http://www.sephora.com
Shiseido: http://www.shiseido.com
Vincent Longo: http://www.vincentlongo.com
Vivienne Tam: http://www.viviennetam.com
Zegna Soft: http://www.zegna.com

NOTES

HOSPITALITY FACILITIES

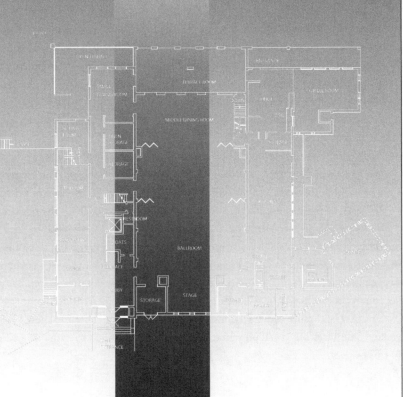

CHAPTER 15

Chippewa Country Club

OVTCOME SVMMARY

The purpose of this exercise is to gain proficiency in writing proposals and contracts and investigating resources for the same. It is divided into two submittal requirement parts. The first part covers preparing a proposal for the project. The second part encompasses writing a design contract in the form of a letter of agreement.

client profile

The board of directors of Chippewa Country Club (CCC) has requested a proposal and letter of agreement from you to develop a master plan to completely renovate and refurbish the interior of the clubhouse. The client is knowledgeable regarding the design process, because there are several industry professionals at the board level. However, be mindful that decisions made by an

Figure 15–1 Artist's sketch of clubhouse exterior (1930)

entity, such as a board, may be more time consuming than those made by a typical client. You will be partnering with an architectural firm that is also part of the design team. The proposed architectural building changes can be found on the CD, in the drawing labeled chippewa.dwg.

In order to understand the goals and objectives of the club members, you must prepare a questionnaire. This anonymous survey will be distributed to the members by the board of directors. The board will use the survey to ensure that the future direction of the club is determined according to the wishes of the membership. The fee to develop this survey is to be included in your letter of agreement as an additional service because the board is receiving proposals for it from others as well.

The renovation would be completed in three phases, beginning with the First Level, the Upper Level, and then the Lower Level. The club must remain open during renovations and it is projected that the work can be completed in 3 to 5 years. They have not determined if they will be using capital funds, loan funding, or member assessment for the project costs.

Your agreement is to reflect two fee options for the interior of the club: First, to master plan and design the entire club and provide complete interior construction documents at one time. Second, to master plan and design the entire club and provide interior construction documents for each level as three separate interior construction document packages, one each year. The purchasing of the goods will be sent out to bid.

The proposal must address the needs and goals of the client. Keep these key issues in mind. Make it an interesting visual presentation and tell your story as to why you are best qualified to do this project.

CCC was established in 1911 and the clubhouse was built in the same year. The country club has 360 full-time golf memberships. It has undergone many minor remodels over the 90 plus years of its existence. Currently, there is no continuity to the plan or design of the interior space. Design throughout is to remain consistent with the building style.

The renovation of the interior of the clubhouse should strive to function well for the members with an aesthetic based on the ideals of neatness, discretion, quality, and consistency. They have requested an elegant design for the new interiors with charm, warmth, and stability, and above all, one that provides convenience and comfort for the present and future of the clubhouse. The end goal is the creation of a unique blend of age, vitality, continuity, and function. They desire solutions and qualities that have a long life and give high levels of satisfaction over that long life. The clubhouse should symbolize the character of a country club and also help people make the choice about what kind of country club they want to belong to. The final design solution must put the clubhouse to work in economically viable ways and make it a better place for members and guests to enjoy.

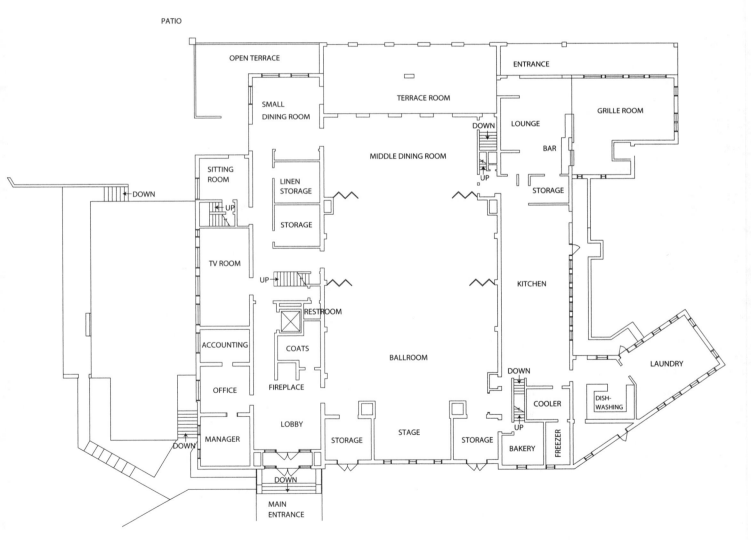

Figure 15–2 First Level Existing Plan

project details

General and Architectural Information

Location: Chippewa Country Club (assume that it is located in your community)

Gross square footage: Three stories, 24,100. First Level: 10,000. Upper Level: 5,000. Lower Level: 9,100.

Project budget: Not established. Bidding is required.

Existing construction renovation. Current facility plans and specifications are not available. Site information gathering is required.

Consultants required: Architect, structural engineer, electrical engineer, and mechanical engineer

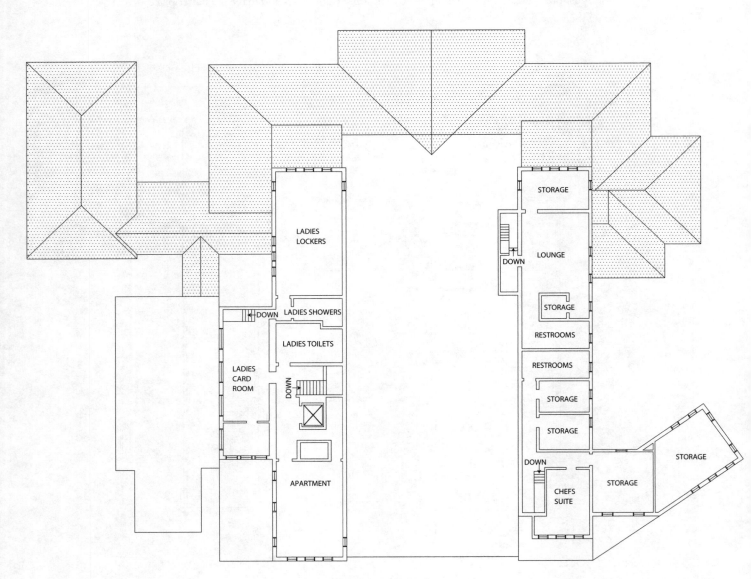

Figure 15–3 Upper Level Existing Plan

Codes and Governing Regulations

Occupancy: A-2 Assembly (+300) and B Business

Comply with the most current basic requirements of the International Building Code (IBC) and the Americans with Disabilities Act.

Facilities Required

First Level: Lobby, Cloakroom, Men's Restroom, Ladies Restroom, Formal Dining Room, Mixed Grill and Bar, Prefunction Room/Private Dining, Ballroom, and Patios

Upper Level: Ladies Locker Room; Ladies Showers; Ladies Toilet; Card Room; Offices of Manager, Assistant Manager, Controller, and Secretary; Employee Lounge; and Restrooms

Lower Level: Men's Locker Room, Men's Showers, Men's Toilet Room, Men's Grille Room, Shoe Room, Pool Decks, Men's and Ladies Pool Showers, Snack Bar, and Fitness Room

The Lobby is to include a formal reception desk and seating arrangement. The Formal Dining Room is to seat 55 to 60 persons, and the Mixed Grill and Bar is to seat 60. The Ballroom is to seat 380, and 220 with a dance floor. Patios will function as outdoor dining as well. The Men's Grille Room and bar is to seat 60.

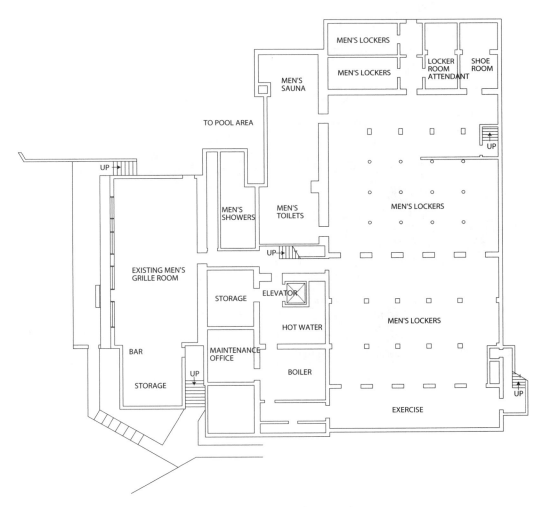

Figure 15–4 Lower Level Existing Plan

SUBMITTAL REQUIREMENTS

Proposal   Date Due_____

   A. Your proposal should include a cover page, title page, table of contents, analysis of problem and statement of concept, pictorial concepts, project methodology, staffing and schedule, interior costs, design firm information, and references.[1]

Letter of Agreement   Date Due:_____

   A. You may use a standard contract or draft one of your own using a fixed fee method (also referred to as a flat fee, stipulated sum or lump sum method).

   B. Your letter of agreement should include dates, identity of client and designer, project description, scope of services, fees, terms, reimbursable costs and additional services, time limit of agreement, party responsibilities, ownership and use of documents, assignment, arbitration and termination, retainer, and signatures.

   C. Submit the above on 8½″ × 11″ white bond paper and bind. The binder cover is to be identified in the lower right-hand corner with your name, the date, the project name and number, the instructor's name, and the course title.

Evaluation Document: Design Agreements and Proposals Evaluation Form

Project Time Management Schedule: Schedule of Activities, Chapter 31.

The AutoCAD Lt 2002 drawing file name is chippewa.dwg and can be found on the CD.

REFERENCES

Books

Ballast, David Kent. 2002. *Interior design reference manual.* 2d ed. Belmont, Calif.: Professional.

Piotrowski, Christine M. 2002. *Professional practice for interior designers.* 3rd ed. New York: Wiley.

Special Resources

Various useful forms and samples can be found in the above-referenced textbook by Christine Piotrowski on pp. 258–260, 272, 273, 275–289, 293, and 386.

Additional resources are in *Designing Interiors.*

American Institute of Architects: http://www.aia.org

American Society of Interior Designers: http://www.asid.org

International Interior Design Association: http://www.iida.com

NOTES

_____

_____

_____

1. Piotrowski, Christine M. 2002. *Professional practice for interior designers.* 3rd ed. New York: Wiley.

CHAPTER 16

Royal Prince Hotel

OUTCOME SUMMARY

The project focuses on developing conceptual skills for the design of hospitality and restaurant facilities and blending Western and Eastern cultures. The project features two phases of the design process: identifying and analyzing the client's needs and goals, and developing conceptual skills through schematic or initial design concepts. You will be working in metric scale and applying the principles of Feng Shui to the interior of the space and applying situational research techniques.

client profile

This project is a joint venture between you and the architects. To some extent it is a test for future work with the client. Your responsibility is to develop several planning and design concepts (rough space planning and concept sketches) for the first and second levels of a commercial hotel that almost exclusively accommodates business clientele in Taiwan. The project involves the first two stories of the seventeen-story hotel, including the lobby and one restaurant space that includes a Wine Bar, Bistro Dining, and fine tableside service.

Michael, the owner, wants the North American or European business traveler to feel comfortable in an elegant background. He wants to stay away from a stark rectilinear look and prefers a Western or European look. His vision includes the use of rich materials such as marble, wood, slate, brass, and glass. In designing the public areas, project a personality of the hotel that is both aesthetically pleasing and highly functional. The furnishings, materials, and finishes must be appropriate to the concept of the hotel as well as durable and easy to maintain.

The owner has hired a Feng Shui consultant in Taiwan as part of the project team who will be reviewing the design of the hotel. He has directed you to apply Feng Shui to the substance and form of the space in order to create harmony and an even flow of ch'i throughout. Environmental conditions, directional orientation, spatial relationships, form, shape, color, and intent are all design elements quite familiar to you. Now consider these factors to bring the overall environment into alignment with the natural laws of flow.

In planning these facilities it is important to consider flow, expanse, volume of business, tempo of service, direction, the ADA, and movement throughout areas. Pay attention to the psychology of visual space, acoustics, olfactory space, and lighting levels. Use natural and artificial lighting elements and sources for general, task, and accent lighting.

The Royal Prince Hotel is located in an urban center and will offer well-appointed accommodations and unrivaled service and style to the business traveler. With a location that provides easy access to the airport, harbor, and industrial parks, it is quite convenient for travelers in the area. In addition, each of the planned roomy, 168 guest suites feature abundant natural light, individual climate control, Internet Speedway, satellite and pay movie channels, electronic key cards, in-room safe deposit boxes, dual-line IDD telephones with voice mail, fax, and PC modem connections.

Business needs will be accommodated fully with an executive business center, accessible 24 hours a day, the latest multimedia equipment, and catering from the restaurants. Meeting rooms will provide seating for up to 200 as well as private rooms for smaller assemblies. The hotel will provide individualized assistance, including a personal secretary, travel consultant, and chauffeur.

Royal Prince's variety of fine restaurants will include the Ocean Reef, where panoramic vistas will accompany fresh seafood. The Mirage, Le Chateau, and Yangtse River restaurants will serve the finest in French, Continental, and Shanghainese cuisine. And to top it off, there is a salon planned to stimulate the mind and soul, complete with sauna, steam baths, and works of art presented in scheduled exhibitions.

project details
General and Architectural Information

Location: Royal Prince Hotel, Taiwan, R.O.C., Seventeen stories

Gross square footage: First Level: 654 square meters. Second Level: 1,026 square meters (approximately 11,700 square feet)

New construction. Preliminary architectural plans are available. The architect will administer construction documents and the bidding of the interior fixture and finish work.

Consultants required: Architect, structural engineer, electrical engineer, and mechanical engineer

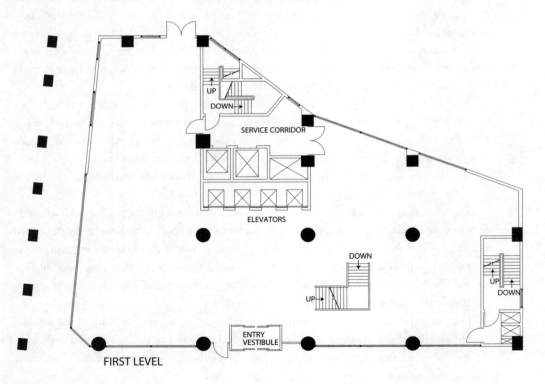

Figure 16–1   Floor Plan: Lobby Level

Codes and Governing Regulations

The Taiwan Building Code (TBC) is a modern code that is based on a combination of the Uniform Building Code (UBC) used in the western United States, Japan Building Code, and local practice.[1] Localities adopt somewhat different standards based on their geological characteristics. Like California, Taiwan is an earthquake region, and special seismic life safety issues apply to construction. The UBC contains information from the seismological and structural engineering communities to reduce the risk of damage from earthquakes and windstorms.

Occupancy: Classified as a mixed use. R1 Residential, Hotel and A3 Assembly, Restaurants

Building Type: Type I, noncombustible. Fire Sprinkler Suppression System

Comply with the most current basic requirements of the ICBO Uniform Building Code, fire and safety codes, and the Americans with Disabilities Act.

Facilities Required

First Level: Entrance Vestibule, Lobby, Executive Concierge, Registration, Front Office, Front Office Manager, Valet and Bell Captain Room, and Quiet Bar Lounge

Second Level: Elevator Vestibule, Maitre de Host Stand, Lounge, Wine Bar, Bistro Dining, Fine Dining

First Level Spatial Requirements

ENTRANCE VESTIBULE

Planning

- The Entrance Vestibule is located on the plan and is a glass-enclosed area.
- Porters and valets will be on hand outside the front entrance to open doors and call for service of parking cars and luggage handling with the bell captain and valet.
- The bell captain is stationed in the Valet and Bell Captain Room at the rear entrance.

Furniture

- Porter podium
- Ash and waste receptacles and green gardens

Equipment

- Luggage carts
- Telephone and house phone

Design

- Select floor covering, ceiling design treatments, finishes, and materials.
- Provide selection of trims and floor bases.

LOBBY

Planning

- Adjacent to Registration, elevators, and Entrance Vestibule
- Plan to manage the traffic patterns from entrances, corridors, and elevators or stairs. Allow sufficient space for baggage handling.
- Socially, the lobby will also function as a gathering and meeting place for the guests.
- Plan seating areas for twelve.

---

1. A Report on Building Performance, Ji-Ji Taiwan Earthquake, 222 Sutter Street, Suite 300, San Francisco CA 94108. Stan M. Wu, John D. Meyer, I. James Chen, Simpson Gumpertz & Heger Consulting Engineers

- Provide a writing desk and house phone station for guests.
- The open stair will be a major circulation stair to the dining level, which will relieve the elevator loads.

Furniture

- Seating and related furniture, including sofas and lounge chairs
- Case goods
- Accessories, lamps, artwork, area carpets, and plants
- Events directory

Equipment

- Pay phones and house phones
- Fire extinguishers
- Water fountain

Design

- The lobby creates the first impression on the guests. It is small and should feel private and elegant.
- The open staircase winds down into the lobby, making it a strong visual element of the lobby space.
- Select durable materials, suitable for public usages, that have been treated for flammability resistance.
- Design a treatment for the elevator door fronts and the interior of the elevator.
- Address the overall signage philosophy to direct guests to other destinations in the hotel. Select and locate directories and mounted signs. Directory type (event board, monitor, etc.) is your decision.
- Select floor covering, wall and ceiling design treatments, finishes, and materials.
- Provide selection of interior doors, trims, and bases.

EXECUTIVE CONCIERGE

Planning

- Adjacency: Registration Desk and Lobby
- The concierge will provide assistance to guests for activities, such as dinner recommendations and reservations, theater or sporting event tickets, transportation, and leisure or recreation.
- Plan to display magazines, literature, and tour brochures, and service of gourmet coffee and a bowl of fresh seasonal fruit.

Furniture

- Elegant desk and credenza, including ample work surface and storage
- Desk chair
- Two guest chairs
- Task lighting and accessories

Equipment

- Telephone/house phone
- Computer

Design

- Relate to lobby elements.

REGISTRATION

Planning

- Adjacency: Lobby and Office
- Plan for easy access and visibility for arriving guests. Clerks will oversee the guests' arrival and departure, and, therefore, must have visual control of the entrance and lobby area. Clerks will include a registrar and a cashier.
- The area will function as guest check-in and checkout. Provide for services offered, such as safety deposit boxes, key boxes, and mail and message boxes and rate cards.
- Plan for two 6′ computer workstations to perform registration checkout and cashier functions. The cashier will function between the two stations.
- Section of the counter must meet ADA guidelines.
- Provide adequate queuing space in front of the registration counter.

Furniture

- Events directory
- Accessories, artwork, and plants
- Time displays for different time zones

Equipment

- Computer equipment for telephones, voice messaging, registration, and payments
- Three printers
- Fax machine
- Security and fire panel

Design

- Design a highly visual feature focal element behind the counter on the wall.
- Develop a signage concept to identify the area and different functions at the counter.
- Select floor covering, wall and ceiling design treatments, finishes, materials, and fixtures.
- Provide selection of interior doors, trims, and bases.

FRONT OFFICE

Planning

- Adjacency: Registration and Valet/Bell Captain Room
- Front Office must include workstations for the reception/executive assistant, reservations clerk, and switchboard operator.
- Switchboard must be located close to the registration desk.

Furniture

- Workstations must include desk-height work surfaces, including a retractable keyboard tray and pencil drawer, two legal size file drawers, one pedestal with a file and two box drawers with locks, storage for computer disks and CDs, stationery, forms, multiple-size envelopes, and task chair (without arms) with casters for carpet
- Six 36″ wide lateral file drawers with locks
- 4 linear feet of shelving for book and manual storage
- Advanced reservation rack

Equipment

- Shared typewriter
- Computers with printers
- Fax machine

- Copy machine
- Telephone with an intercom

FRONT OFFICE MANAGER

Planning

- Adjacency: Administrative Assistant, Registration
- This is a private office for the manager who is responsible for the supervision of front desk personnel.

Furniture

- Desk-height work surfaces, including a retractable keyboard tray and pencil drawer
- Four 36″ wide lateral file drawers with locks
- Four legal size file drawers
- One pedestal with a file and two box drawers with locks
- Overhead storage of at least 72 linear inches
- 20 linear feet of shelving for book and manual storage
- A management level chair with casters for carpet
- Tack surfaces are required
- Task light at the desk
- Guest seating for one
- Waste receptacle, paper trays, message holder, day calendar holder, artwork, and plants

Equipment

- A computer with a printer
- Telephone with an intercom
- Safe

VALET/BELL CAPTAIN ROOM

Planning

- Adjacency: Front Office and rear entrance
- Plan a locked luggage storage area. Luggage carts will be stored outside, at the rear exit area. Plan locked auto key storage as well.
- Provide a counter for the bell captain and valet.

Furniture

- Counter stool
- Two ticket racks—one for valet and one for luggage

Equipment

- Luggage carts
- Telephone/house phone

Design

- Design a counter fixture.
- Provide locking key storage fixture.
- Provide a locking luggage storage area.
- Select floor covering, wall and ceiling design treatments, finishes, materials, and fixtures.
- Provide selection of interior doors, trims, and bases.

QUIET BAR LOUNGE

Planning

- Adjacency: Lobby, in an open setting
- Provide good visibility.
- Plan a cocktail lounge area that will provide beverage and appetizer food service. It will be used for socializing and casual meetings. Waitress service will be provided.
- Plan a bar with seating for ten to twelve at the front bar. The back bar must include storage for and display of wines and liquors. Storage for twelve types of glassware is also required.
- The under bar is the main work area for the bartender. It must include sinks and space for cleaning, sanitizing, and draining glassware. Provide storage, ice bins, soda-dispensing guns, beer taps, and speed rails. Provide cold mixer, beer, and wine storage.
- Service bar/station for the waitstaff is required, including storage for dishes, flatware, soiled dishes, and coffee service.
- Provide a computer terminal cash register behind the bar and at the service station for dining.
- Seat approximately thirty in the lounge area. Provide furniture groupings for two to six people.
- Provide an entertainment area with a piano.

Furniture

- Bar stools
- Comfortable seating arrangements, including sofa groupings and cocktail tables and chairs
- Plants, accessories, artwork, and table lamps or candles

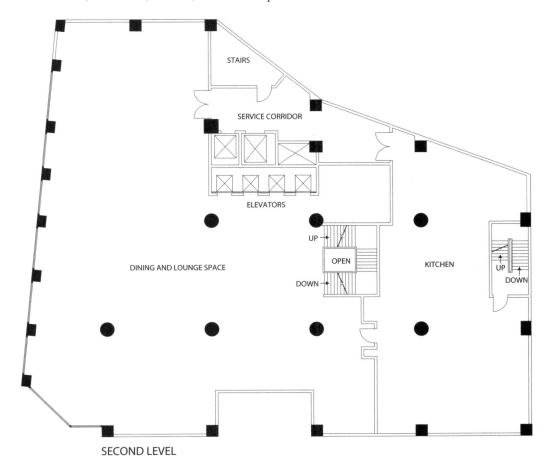

Figure 16–2   Floor Plan: Dining Level

Equipment

- Piano
- Television
- House phones

Design

- Design a distinctive theme.
- Develop a subdued lighting scheme including accent lighting.
- Select floor covering, wall and ceiling design treatments, finishes, and materials that will encourage guests to linger with low sound levels and soft surfaces.
- Provide selection of trims and floor base.

Second Level Spatial Requirements

Elevator Vestibule

Planning

- Adjacency: Lounge and dining facilities
- This is the main transition space from other levels. Allow adequate circulation space for elevators and stairs.
- Provide a coat storage area with coat check counter, and pay phone area.

Furniture

- Any furnishings that you find appropriate for the space

Design

- It should have a focal point, be identifiable, and inviting because it is the entrance to the Lounge and dining facilities.
- Select floor covering, wall and ceiling design treatments, finishes, and materials.
- Provide selection of interior doors, trims, and floor base.

Equipment

- Fire extinguishers
- Water fountain

Maitre de Host Stand

Planning

- Adjacency: Dining entry. Location should be easy to find.
- The station may be a custom or manufactured item. Include a surface for the reservation book, table layouts, task light, telephone, and drawer.

Lounge

Planning

- Adjacency: Elevator Vestibule and dining facilities
- Plan a cocktail lounge in an enclosed area that will provide beverage service, appetizer food service, and entertainment. Waitress service will be provided.
- This area is to be planned as a living room setting. Guests will use it as an area to wait before being seated for dining or simply as a socializing lounge. Seat approximately thirty.

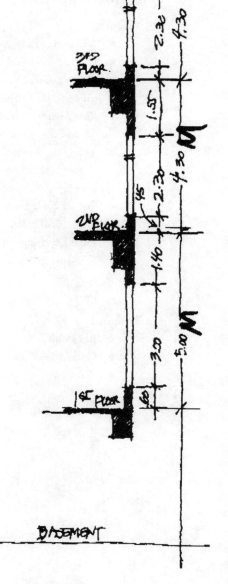

Figure 16–3 Building Section

- Provide furniture groupings for two to six people.
- Provide an entertainment area with piano and small dance floor.

Furniture

- Comfortable seating arrangements, including sofa groupings and cocktail tables and lounge chairs
- Accessories, lamps, artwork, table lamps or candles, and plants

Equipment

- Piano
- Sound system

Design

- Provide theater lighting for entertainment area and dance floor.
- Select floor covering, wall and ceiling design treatments, finishes, and materials.
- Provide selection of interior doors, trims, and floor base.

WINE BAR

Planning

- Adjacency: Bistro Dining and Lounge
- This will be the main focal point of the Bistro Dining area, where the largest part of the wine inventory will be maintained.
- Include a full bar with wine displays.
- Seat eight to ten at the front bar. Back bar must include storage for and display of wines and liquors. Storage for twelve types of glassware is also required.
- The under bar is the main work area for the bartender. It must include sinks and space for cleaning, sanitizing, and draining glassware. Provide storage, ice bins, soda-dispensing guns, beer taps, and speed rails. Provide storage for cold mixers, beer, and wine.
- Provide a service bar/station for the waitstaff.
- Provide a computer terminal cash register behind the bar and at the service station for dining.

Furniture

- Bar stools
- Plants, accessories, artwork, and table lamps or candles

Design

- Coordinate with the design concept of adjacent areas.
- Provide more subdued lighting than that in the dining areas.
- Select floor covering, wall and ceiling design treatments, finishes, and materials.
- Provide selection of interior doors, trims, and floor base.

BISTRO DINING

Planning

- Adjacency: Wine Bar and Fine Dining
- Service includes buffet breakfast service, lunch service, and candlelight dinners. The Bistro Dining area will have a higher turnover than Fine Dining.
- Plan for table and banquette seating to accommodate approximately fifty persons, with a combination of booths banquettes, chairs, and tables.
- Provide two service stations that include storage for dishes, glasses, flatware, soiled dishes, and coffee service.

Furniture

- Dining tables and chairs and banquettes
- A variety of accessible seating
- Plants, accessories, artwork, and table lamps or candles

Design

- Lighting is to be on dimming system: brighter lighting for breakfast and lunch, and subdued lighting for dinner service. Consider visual comfort, safety, mood, and ability to read menus and enhance the appearance of food.
- Select colors that complement the interior scheme and enhance the appetite.
- Select floor covering, wall and ceiling design treatments, finishes, and materials. Finishes must be easily maintained, cleanable, resistant to grease, spills, abrasion, snags and pilling, and consider safety (falls).
- Provide window treatments.
- Provide selection of interior doors, trims, and floor base.
- Include design of tableware, linens, and glassware as well as graphic design of menu cover, coasters, and matchbooks.

FINE DINING

Planning

- Adjacency: Bistro Dining and Lobby Vestibule
- Service will include full-service lunch and candlelight dinners.
- Seat approximately 75, with a combination of booths, chairs, and tables.
- The traffic should flow easily from the entry throughout the dining areas. Traffic from the dining room to the kitchen must be efficient. Provide main circulation aisles from 3′ to 5′ wide. Provide activity space for serving that does not interfere with the circulation space.
- Provide a waitress station with computer terminal for order entry and payment of bills.
- Consider sight lines and keep stations and interiors of the kitchen out of dining room views.
- Plan an area that can be set up as a breakfast bar, salad bar, or food buffet. Fixture counter should be accessible. Include steel food equipment with drains, serving platters, condiments, and dinner plates. Provide for a large food visual presentation.

Furniture

- Any combination of booths, chairs, and dining tables
- Larger tabletops and more comfortable chairs
- A variety of accessible seating

Design

- Design as an upscale gourmet restaurant that is slightly formal and very classy.
- Provide good acoustical control to keep noise levels low.
- Select floor covering, wall and ceiling design treatments, finishes, and materials.
- Provide selection of interior doors, trims, and floor base.
- Include design of tableware, linens, and glassware as well as graphic design of menu cover, coasters, and matchbooks.

SUBMITTAL REQUIREMENTS

Phase I: Programming   Date Due:_____

A. To better understand how hotels and restaurants function, apply the technique of situational research. Visit these facility types and monitor the actions of both the customers and the service providers. Tour various hospitality facilities and observe how the planning and design requirements are addressed. Have a class discussion, sharing the occurrence and your observations.

B. Preliminary plan studies of the first and second levels must include lobbies, the restaurant, lounges, and back areas. Submit a scaled freehand final plan layout of furnishings and fixtures and ceiling design. Plans should be presented in 1:100 metric scale.

Evaluation Document: Programming Evaluation Form.

Phase II: Conceptual Design   Date Due:_____

A. Perspective line drawing of the Lobby level must indicate the interior design character of the space. Submit as a black line drawing on white bond paper. Provide an additional copy of floor plans with materials indicated by description. Plans should be presented in 1:100 metric scale.

B. Prepare materials, colors, photos, and sketches that include the design of tableware, linens, glassware, and the graphic design of menu cover, coasters, and matchbooks.

C. Submit unmounted color and materials and cut sheets of proposed furnishings. Submission should include project record sheets identifying materials and their general locations.

Evaluation Document: Conceptual Design Evaluation Form

Project Record Sheets: See Chapter 31, Project Record Sheets.

Project Time Management Schedule: See Chapter 31, Schedule of Activities.

The AutoCAD LT 2002 drawing file name is prince.dwg and can be found on the CD.

REFERENCES

Books

Baraban, Regina S., and Joseph F. Durocher. 2001. *Successful restaurant design.* 2d ed. New York: Wiley.

Curtis, Eleanor. 2001. *Hotel: Interior structures.* New York: Wiley.

Wong, Eva. 1996. *Feng Shui, the ancient wisdom of harmonious living for modern times.* Boston: Shambhala.

Wong, Eva. 2001. *A master course in Feng Shui.* Boston: Shambhala.

Trade Magazines

Hospitality Design Magazine: http://www.hdmag.com

Interior Design Buyers Guide: http://www.interiordesign.net

Product Manufacturers

The Amtico Company: http://www.amtico.com
B&B Italia: http://www.bebitalia.it
Bentley Mills: http://www.bentleyprincestreet.com
Brayton International: http://www.brayton.com
Cascade Coil Drapery: http://www.cascadecoil.com
Cassina: http://www.cassina.it
Cassina USA: http://www.cassinausa.com
Chairmasters: http://www.chairmasters.com
Deepa Textiles: http://www.deepa.com
Donghia: http://www.donghia.com
Falcon Products: http://www.falconproducts.com
Invision Carpet Systems: http://www.invisioncarpet.com
Jhane Barnes Textiles: http://www.jhanebarnes.com
Knoll Textiles: http://www.knoll.com
Loewenstein: http://www.loewensteininc.com
Maharam: http://www.maharam.com
Metro Furniture: http://www.metrofurniture.com
MoDAVATION: http://www.modavation.com
Pallas Textiles: http://www.pallastextiles.com
Unika Vaev Textiles: http://www.unikavaev.com
Vanceva: http://www.vanceva.com

CHAPTER 17

The Verniccini Oven

OUTCOME SUMMARY

The project focuses on applying the design process to a restaurant franchise and on the generation of multiple design ideas, including practice in graphic visualization techniques. Requirements are developing a schematic, or initial prototype design, including conceptual design sketches, to establish a visual concept. These ideas will be communicated through storyboards, drawings, sketches, material samples, and very loose space plans.

client profile

Fosca Verniccini owns the ten-year-old restaurant that is expanding rapidly. She wants an identifiable concept and signature image before she proceeds with plans to franchise the operation. Distinctive Italian food at reasonable prices is Verniccini's mission, and the concept of uniting food preparation with eating is her vision. The restaurant has a full bar, and various locations could include outdoor dining.

Upon entering the current restaurant you are greeted by the hostess in a space filled with Fosca's signature "survival kits," which include an array of pasta, prepared foods, packaged foods, and baked goods. Shelves filled with rows and rows of pasta, cans, bottles, and jars of imported specialty goods; and of course pane; lots of pane.

The informal, child-friendly dining room is currently designed to seat approximately 150 diners. The authentic, wood-fired, brick pizza oven is in an exhibition kitchen where the specialties are prepared by the chefs. It is the focal point of the open dining area. The gourmet pizzas are prepared in full view of the diners as part of the entertainment of the evening. Displayed in the glass along the counter are the colorful presentations of cheeses, tomatoes, meats, vegetables, and mountains of flour and dough. The dough is freshly pressed and kneaded and the tossing is the featured attraction. A self-service salad bar where diners can select antipasto salads, fresh fruit, vegetables, and assorted delicacies is also a featured attraction, and the presentation of food is the focal point of the salad bar. It is hard to resist the enticing aroma coming from the authentic, wood-fired, brick pizza oven. It may very well be your starting point for the whole concept. It is all theater and needs a great staging setting.

The locals who are in the know come not just for the scene but also for the roster of of authentic Italian cuisine. The menu is organized as a bona fide Tuscan Tratoria bill of fare made up of antipasto, first plate, and second plate, etc. Starters are Tuscan Bruschetta and Calamari or the homemade pane. A treat of extra virgin olive oil and imported cheese are sitting on each table. For pizza dishes, to name a few, there is Portabello Pizza with grilled marinated portabello mushrooms and Roma tomatoes topped with mozzarella and goat cheese. The Al Pomodori, a thin-crust pizza with Roma tomatoes, basil, mozzarella, red onions, and sun-dried tomatoes, is an all-time favorite of the regulars. The menu offerings also include rotisserie grilled meats and fowl and mesquite-grilled fresh fish.

project details

General and Architectural Information

Location: Verniccini Oven, Peachtree Town Center, Atlanta, Georgia

Gross square footage: 3,500. Dining square footage: 2,250

Hours of operation: Generally, Monday –Friday: 11:30 a.m. to 10:30 p.m.; Saturday: 11 a.m. to 10:30 p.m.; Sunday: 11 a.m. to 6 p.m. Weekend hours could vary depending on location.

Interior design prototype. Current facility plans and specifications are not available. Plan for locations, including malls, neighborhood centers, airports, and stand-alone buildings. A suggested plan and elevation follows.

Consultants required: Architect, electrical engineer, and mechanical engineer

Codes and Governing Regulations

Occupancy Classification: A2 Assembly

Construction Type: Type I, Noncombustible. Sprinkler Fire Suppression SystemFollow the most current version of the International Building Code (IBC), the International Fire Code (IFC), and ADA guidelines for restaurant spaces. A minimum of two exits is required.

Facilities Required

Storefront and Entry Vestibule, Retail Shop and Cashier, Hostess Station, and Dining, including Pizza Kitchen and Bar Area. (Back of house and restrooms are not part of this exercise.)

Spatial Requirements

STOREFRONT AND ENTRY VESTIBULE

- A windowed storefront (a minimum glass window area of 50% of the storefront) with swinging entry doors is preferred. Keep in mind that the ceiling is a visible element of the interior, and, therefore, should be treated with much care. Lighting fixtures should enhance the special character of the space. A sample storefront area is provided in Figure 17–2. It is not mandatory that you use it.
- Sign may not exceed 60% of the storefront width, with a maximum letter height of 24″.
- A menu board plaque must be located and lit from a shielded light source.
- Establish an identifiable inviting look that is functional and addresses circulation.
- Provide for coat storage.

RETAIL SHOP AND CASHIER

- Adjacency: Entry Vestibule and Hostess Station
- Create an attention-getting front display that will gesture you inside and entice the appetite.
- Provide fixtures for the "survival kits," pasta, prepared foods, packaged foods, cans, bottles and jars of imported specialty goods, and baked goods.
- Cashier station must include a service counter with point of sale (POS) station, and a visual presentation of merchandise on the back counter and wall.

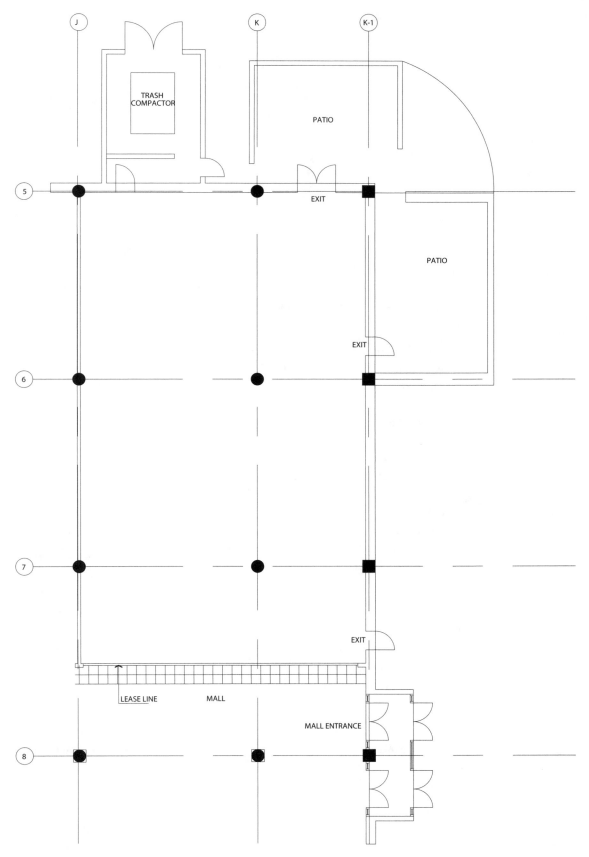

Figure 17–1   Floor Plan

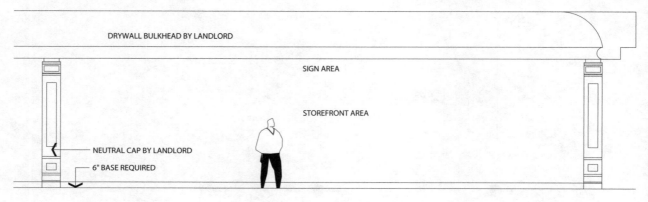

DRYWALL BULKHEAD BY LANDLORD

SIGN AREA

STOREFRONT AREA

NEUTRAL CAP BY LANDLORD

6" BASE REQUIRED

Figure 17–2 Storefront Elevation Criteria

HOSTESS STATION

- Adjacency: Dining entry and Cashier. Location should be easy to find.
- The station may be a custom or manufactured item, including a surface for the reservation book, table layouts, task light, telephone, and drawer.

DINING

- The traffic should flow easily from retail throughout dining. Traffic from the dining room to the kitchen must be efficient. Provide main circulation aisles from 3′ to 5′ wide. Provide activity space for serving that does not interfere with the circulation space.
- Consider sight lines that include views of the oven and exhibition kitchen interior. Place food displays in prominent locations. Carefully consider placement of service stations so they are not part of any prominent sight line or view.
- Plan for table service and bar counter service to accommodate 150 persons, with a combination of booths, banquettes, chairs and tables, and counters and stools. The bar counter service includes beverages and food. Locate the counter to face the Pizza Kitchen.
- Include service stations with storage for dishes, glasses and flatware, soiled dishes, coffee service, and a computer terminal for order entry.
- Provide for a self-service salad bar. This is a featured attraction and the presentation of food is a focal point of the salad bar.

PIZZA KITCHEN

- This is the main focal point of the dining area. The pizza oven and food preparation must be behind a glass enclosure.
- Provide seating at the front bar counter with a view of the Pizza Kitchen, and plan for a fully functioning food and beverage bar.
- Code requires an easily cleanable ceiling.

OUTDOOR PATIO

- Settees, chairs, tables, umbrellas, and planting pots are required.
- Create a distinctive outdoor design theme that may include canopies, trellis panels, planting, fountains, or umbrellas. Enclose the area with a decorative, low height railing.
- Provide fabric-upholstered seat cushions that are suitable for outdoors.

GRAPHICS

- Develop a graphic design for logo, menu cover, coasters, napkins, matchbooks, and shopping bags.
- Select tableware, linens, glassware, flatware, table accessories, and dress style of the waitstaff.

SVBMITTAL REQVIREMENTS

Phase I: Programming   Part A Due: _____

    A. To generate ideas, use the following techniques:

        a. Situation research
        b. Brainstorming
        c. Graphic visualization techniques

Phase II: Conceptual Design   Part A Due: _____   Part B Due: _____

Part C Due: _____   Part D Due: _____   Part E Due: _____

    A. Develop a freehand preliminary plan, showing all walls, ceilings, built-in fixtures, doors, and windows. Label rooms and indicate the square footage.

    B. Develop conceptual ceiling and lighting studies, showing all ceiling conditions and heights.

    C. Develop a conceptual color, material, and furniture scheme.

    D. Develop the graphics package.

    E. Develop several conceptual freehand perspective sketches of the interior, and apply partial color to the perspective and elevations.

You have total freedom to select the medium and methods for the presentation, but you must draw and present several concepts that successfully communicate your ideas in a graphic format. Include plans, storyboards, samples, catalog cuts, and colored sketches in your presentation.

Class evaluation by means of the critique method is required for this exercise using elements from Evaluation Documents Programming Evaluation Form and Conceptual Design Evaluation Form.

Project Time Management Schedule: Schedule of Activities, Chapter 31.

The AutoCAD LT 2002 drawing file name is verniccini.dwg and can be found on the CD. You may substitute any plan or elevation of your choosing but it must have the required square footage.

REFERENCES

Books

Baraban, Regina S., and Joseph F. Durocher. 2001. *Successful restaurant design, 2d ed.* New York: Wiley.

Birchfield, John C. 1988. *Design and layout of foodservice facilities and NREF workbook package.* National Restaurant Association Educational Foundation. New York: Wiley. Farrelly, Lorraine. 2003. *Bar and restaurant interior structures.* New York: Wiley. Special Resources

Hospitality Design Magazine: http://www.hdmag.com

National Restaurant Association and Restaurants USA Magazine: http://www.restaurant.org

NOTES

RECREATION FACILITIES

CHAPTER 18

Metro Club

OUTCOME SUMMARY

The project features a team approach to two phases of the design process: to define team responsibilities, discuss goals, and explore client requirements through design team meetings, buzz sessions, dialogue groups, situation research, and interviews; and to establish optimal space plans using adjacency diagrams, which show the proximity of the various areas and activities. The team is responsible for researching the local codes for health club facilities, and the project concentrates on the programming phase of the design process.

client profile

Distinction in fine dining, gracious service, lavish holiday member parties, and up-to-date athletics is the buzz around town regarding this project, the Metro Club. Located in a high-rise office building in the heart of the metropolis, the Metro Club was founded to present a matchless venue for business and social entertaining, as well as the capability to improve one's personal fitness.

This private business and sport club will consist of formal and casual dining areas, six meeting rooms, a ballroom with seating for 300, and athletic amenities. Technological enhancements planned include videoconferencing, teleconferencing, and Internet access, accessible in all private meeting rooms. The athletic level will offer yoga, Pilates and Precision Cycling rooms in addition to cardiovascular and strength-training equipment with Cardio Theatre and a sports bar, complete with Internet access.

Your objective is to provide programming and planning services for the athletic level. The Metro Club sports and fitness center facilities are to be designed to enhance the health and well-being of its members. When fully equipped it will offer everything members need to implement their personal fitness and sports or wellness activities in a comfortable, engaging environment. The program and services will add to the experience by providing the right amount of fun, motivation, and professional attention to each member's goals.

The charge of the athletic department is to be a resource core providing state-of-the-art health and fitness programs to the members: utilizing up-to-date equipment; in an inspiring atmosphere; with knowledgeable, kind employee associates.

Ground Zero Equipment, the latest line of resistance equipment using the free motion technology, will be the circuit training exercise equipment. It offers all the benefits of free weights, such as the varied movement patterns are capable, without the lack of control. Also planned are a full compliment of cardiovascular Cybex, and free weight equipment in addition to a lap pool, running track, basketball, racquetball, squash, and walleyball (a rage among the local yuppies) courts. The Metro Club will organize tournaments, leagues, ladders, clinics, and lessons for the members. Also included are programs such as tennis, golf, baseball, and inline skating lessons, and interclub competitions in squash, basketball, and rowing.

The Group Fitness program (offered as free classes) includes the following:

- Beginner and advanced traditional aerobics classes (fifty classes each week), including Low and High Impact, STEP, Sculpt and Stretch
- Specialty classes like Kickboxing Fitness, Resista-Ball, Precision Cycling, Circuit, and Funk. Several classes target specific areas (legs, abs, etc.)
- Mind-body wellness specialty classes such as tai chi, yoga, and Pilates with a designated studio

The athletic services include the following:

- A fitness assessment, to evaluate cardiovascular fitness, flexibility, body fat composition, and strength, which will take approximately an hour. A trainer will make recommendations for improving fitness level safely and effectively. This will be complimentary to new members.
- Certified trainers, to provide motivation and advice for people of all fitness levels. Fitness packs of yoga, Pilates, or personal training will be offered for additional costs. A learner's permit will be available for 14- and 15-year-olds.
- Squash instruction for all levels and group clinics. Racquet stringing and rental will be available.
- Massage therapists, to provide traditional Swedish or sports massage, as well as reflexology and trigger point.
- Specialty and individual sessions will be available in yoga.
- Pilates body conditioning, a low-stress method, to improve posture, strengthen and tone muscles, increase coordination and balance, and streamline the body. This program is challenging to the elite athlete, yet gentle enough for seniors and pregnant women. It will be offered for rehabilitation from injury because it restores balance and mobility to the joints.
- Nutrition counseling with a licensed dietitian, for a dietary analysis.
- Locker and laundry services.
- Shoe-shine service.
- Leagues and tourneys, in which members can compete against each other and against other clubs in basketball, racquetball, squash, handball, volleyball, and walleyball.
- Special events, such as weight loss challenges, fun runs, rowing classes, and self-defense. Special events are to be tailored around member interests.
- Swim lessons packages for adults and children.
- Childcare services: Monday–Friday: 9:30 a.m. to 11:30 a.m.; Monday, Wednesday, and Friday: 5:00 p.m. to 7:30 p.m.; Saturday: 9:00 a.m. to 1:00 p.m.; Sunday: 10:00 a.m. to 1:00 p.m.

Wellness services will include spa therapy, stress management, smoking cessation, back care, eating management, and nutrition classes. The unique sports programs will be designed and developed specifically for the needs of members.

Club locker rooms are to be equipped with comprehensive selection of personal amenities. A permanent locker with the convenience of laundry service for workout clothes is a fee service. Day lockers will also be available. In addition, each locker room will have a steam room, sauna, whirlpool, and a comfortable lounge area. Complimentary facilities include a babysitting room with games and activities for children of all ages.

In addition to the athletic amenities, outstanding cuisine will be offered in the dinign room or club room. Cooking classes with the club's chef as host instructor is one of the planned monthly events. A complete range of private party and business function services, including a personal planner, will provide for all meeting requirements, including meals, for groups of 2 to 200 (seated) and up to 400 standing receptions. You are not required to design these facilities.

The aspiration of the Metro Club is to provide a network of comfortable and inviting venues for members to meet and entertain clients and associates. The aim is to skillfully integrate personalized service and technology within a business-conducive environment, orchestrated to support and leverage the business objectives of the membership. The business facilities, not a part of your project, will include an e-lounge, multimedia-capable meeting and conference rooms, member workstations, document center, videoconferencing, high-speed Internet access, high-tech audiovisual equipment, and concierge services.

Membership classification structure includes the following:

- Executive membership entitles members to privileged service and personal attention. These benefits include dining and social privileges, sports and fitness, racquet sports at no fee, exemption from guest fees, and a complimentary personalized locker with daily laundry service. Co-members of an executive member have the same privileges as a fitness member.
- Fitness members and their co-members enjoy full athletic, dining, and social privileges. The fitness member incurs court fees, guest fees, and locker fees.
- Social members and their co-members have full use of the club's social and dining facilities, but do not have racquet sports or athletic privileges.
- Nonresident members have the same privileges as fitness members but are available to persons whose business and residence are located 50 miles or more from the club.
- Corporate membership programs are customized to meet individual business objectives. They specialize in Corporate Wellness Programs. Corporate membership allows the use of private meeting space at over 210 clubs around the world (one primary corporate designee per corporate membership).

All-inclusive packages:

- Gold: Members must elect Gold for access to golf and athletic privileges at associate country clubs and sports clubs. Green fees are waived with two rounds of golf per membership, per property every 30 days. Also includes specially priced club resort discounts, courtesy professional quality golf clubs, and 50% off accommodations at selected golf resorts, relocation, and legacy benefits.
- Silver Plus: Members must elect Silver Plus to receive access to golf at a 50% discounted green fees at all clubs included on the roster. This benefit includes two rounds of golf per membership, per property every 30 days, athletics, relocation, and legacy privileges.
- Metro Society: This enhancement of membership adds privileges at additional area clubs and an extensive list of concierge services, including VIP tickets to concerts, theater, and sporting events.

project details

General and Architectural Information

Location: Metro Club, Fitness Level Four

Gross square footage: 30,000

Hours of operation: Monday–Friday: 5:30 a.m. to 9:00 p.m.; Saturday: 7:00 a.m. to 5:00 p.m.; Sunday: 9:00 a.m. to 5:00 p.m.

Number of employees: Nine fitness trainers, seven massage therapists, nutritionist, squash pro, golf pro, spa cosmetologists, athletic director, receptionist, four locker room and athletic floor attendants

Opening or move-in date: Soft opening, October 1st. Grand opening, October 17th

Existing construction. Current facility plans and specifications are available. Site information gathering is required.

Consultants required: Architect, electrical engineer, and mechanical engineer

Codes and Governing Regulations

Occupancy: A3 Assembly, Recreation. (BOCA)

Building Type: Type 2A. Protected. (BOCA)

Fire Protection: Fire Sprinkler Suppression System

The project was designed to comply with the current edition of the BOCA National Building Code, the National Electrical Code (NEC), and the Americans with Disabilities Act.

Since the completion of this project, the Pennsylvania Department of Labor and Industry will adopt as its Uniform Construction Code (UCC) the International Building Code (IBC) and most of its referenced standards. Your code research should reflect this adoption.

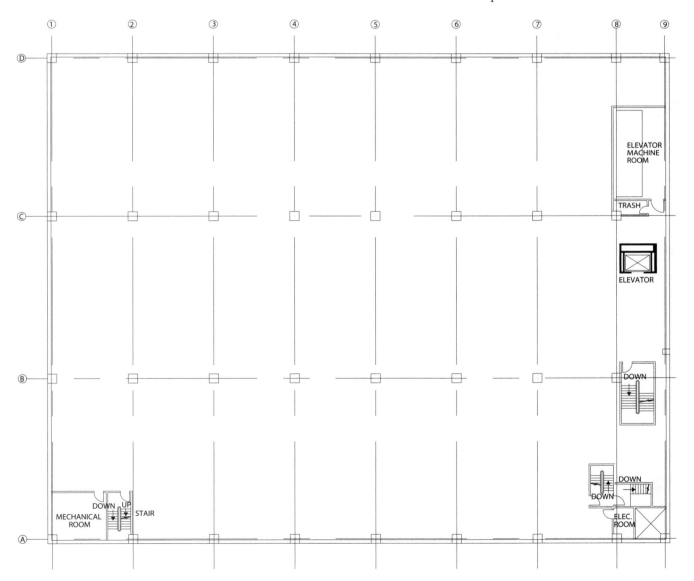

Figure 18–1   Athletic Level Plan.

Facilities Required

See Table 18-1. The men's locker room and athletic staff facilities are located on fitness level five and are not part of the scope of your project.

FACILITIES REQUIRED	APPROX SQUARE FEET
Administration	
Lobby	450
Reception Desk	170
Pro Shop	160
Children's Play Room	300
Children's Toilet Room	40
Dining Court	1000
Athletic Director's Office	150
Physical Therapist's Office	100
Physical Trainer's Office	100
Member Athletic Records Area	100
Women Employees' Locker Room	200
Men Employees' Locker Room	200
Laundry	330
Maintenance Room	150
Electrical Equipment Room	80
Electrical Panel Closet	10
Athletic Floor	
Running Track	2,600
Cardiovascular Equipment	1,300
Weight Machines	1,200
Free Weight Room	700
Racquetball Court—4 required	735
Squash Court	525
Basketball Court	2,130
Specialty Class Room	725
Aerobics Class Room	1,500
Fitness Testing Room	125
Lap Pool	2,300
Pool Equipment Room	140
Pool Shower	10

FACILITIES REQUIRED	APPROX SQUARE FEET
Spa Therapy	
Waiting	110
Therapy Room—2 required	65
Yoga Room	400
Locker Room	80
Men's Dressing Room	40
Ladies Dressing Room	40
Ladies Locker Room	
Lobby	250
Attendant	30
Janitor Closet	30
Massage Room—3 required	85
Sauna	60
Steam	80
Steam equipment	15
Whirlpool	150
Whirlpool equipment	60
Locker Area	2,500
Private Dressing Room—4 required	15
Dressing Area(s)	200
Toilets	200
Showers	400

Table 18–1   Facilities required

SUBMITTAL REQUIREMENTS

Phase I: Programming Part A Due: _____   Part B Due: _____

Part C Due: _____   Part D Due: _____   Part E Due: _____

Part F Due: _____   Part G Due: _____   Part H Due: _____

A. Your instructor will organize the design team and negotiate a compensation method. Verify each team member's ability to meet the time schedule for the project.

B. Given the required facilities and space requirements, research various arrangements and clearances for the above-outlined spaces. Research fixture, furniture, and equipment for the same. Develop a programming form to assist with the rest of the information gathering process.

C. Proceed next with situation research regarding the activities of the facility, and, using your programming forms, interview an athletic director and private athletic club manager. Sample interview questions follow[2]:

   a. Activity requirements for each area or room:
      i. Primary and secondary activities
      ii. Characteristics of the activities
      iii. Frequency of the activity
      iv. Preferred arrangement and location of user or activity area
      v. Who participates in the activity?
      vi. Is a shared area or dedicated area required for the activity?
      vii. Special requirements for the activity: equipment, security, audio, lighting, and acoustics
   b. Adjacencies:
      i. Required communication with others
      ii. Required interchange of materials or equipment
      iii. Level of proximity
      iv. Required zoning of related activities
      v. Required proximity to circulation areas, elevators, stairs, windows, andservice facilities
   c. Furnishings and equipment:
      i. Types of furniture or equipment required
      ii. Sizes of equipment and placement of equipment or furnishings
      iii. Communication equipment: telephones, computers, facsimile machines, modems, etc.
      iv. Storage requirements
      v. Electrical requirements
      vi. Accessories used
      vii. Specific style, quality level, and ergonomic need
      viii. Display space
      ix. Audiovisual equipment requirements

D. Discuss the following mock preliminary plan; check with your instructor before proceeding. Assign a plan check manager from your group who will arrange a meeting for a review of the codes as they apply to life safety and accessibility requirements for your project before proceeding with the planning phase. In the event your local authorities are not able to accommodate a school group request such as this, ask a code official fire marshal or architect to visit your class and share his or her knowledge in reviewing your project.

---

2. Gathereed from numerous design firm resources and authors Kilmer, Piotrowski, and Reznikof.

E. Hold two design team meetings, including a buzz session and dialogue group for the project. In the buzz session, discuss the topic of space planning and propose alternative strategies. In the dialogue group session, share your knowledge from the research you have completed and arrive at a planning and design approach for the project.

F. Analyze the program developed and prepare a programming matrix keyed for adjacency locations. A sample blank matrix is provided in Chapter 31.

G. With the information gathered from the matrix, provide a minimum of three bubble diagrams for the spatial relationships of the areas required.

H. Provide a one-page typed concept statement of your proposed planning strategy and initial design concept approach.

Presentation formats for Parts A through H: White bond paper, 8½″ × 11″, with a border. Insert in a binder or folder. The matrix and bubble diagrams are to be completed using the CADD program.

Evaluation Document: Programming Evaluation Form

Phase II: Conceptual Design

Part A Due: _____ Part B Due: _____ Part C Due: _____

A. Provide a preliminary partition and area location plan, showing all walls, ceilings, built-in fixtures, doors, and windows. Label departments and rooms and indicate the square footage required and the square footage planned. These plans can be freehand drawings or CADD drawings in 1/8″ scale. Drawings must have an appropriate border. Finished drawings must be a blueprint or white bond copy.

B. Develop the above partition plan, including a preliminary equipment, fixture, furniture, and furnishings layout of each area and room. Label areas and rooms and indicate the square footage required and the square footage planned. This plan can be a freehand drawing or CADD drawing in 1/8″ scale. Drawings must have an appropriate border. Finished drawings must be a blueprint or white bond copy.

C. Develop the above plans into a final floor plan, showing all items noted in the program. This plan is to be completed using the CADD program.

Evaluation Document: Conceptual Design Evaluation Form

Project Time Management Schedule: Schedule of Activities, Chapter 31.

The AutoCAD LT 2002 drawing file name is metroclub.dwg and can be found on the CD.

REFERENCES

Books

Panero, Julius. 1989. *Human dimension and interior space: A source book of design reference standards.* Norwell, Mass.: Kluwer Academic.

Scuri, Piera. 1995. *Design of enclosed spaces.* Norwell, Mass.: Kluwer Academic.

Trade Magazines

Club Industry: http://www.clubindustry.com

Fitness Management: http://www.fitnessworld.com

Associations

International Health, Racquet and Sportsclub Association (IHRSA): http://www.ihrsa.org

Medical Fitness Association (MFA): http://www.medicalfitness.org

Special Resources

American Locker Security Systems: http://www.americanlocker.com

Classic woodworking: http://www.classicwoodworking.com

Club Resource Group: http://www.clubresourcegroup.com

Cybex International: http://www.ecybex.com

Equipment Websites: http://cc.ysu.edu/exsci-alumni/equipmentwebsites.htm

Fitness Alliance.net: http://www.fitnessalliance.net

Free Motion Fitness: http://www.iconfitness.com

NOTES

NOTES

RESIDENTIAL FACILITIES

KITCHEN 26N   KITCHEN 26S

APARTMENT
26N   APARTMENT
26S

5   4   6   7
8   7

26TH FLOOR FOYER

METER CLOSET   METER CLOSET
ELECTRICAL CLOSET

2   3

NORTH   SOUTH
ELEVATOR   ELEVATOR

STAIR   STAIR

UP   UP
DOWN   DOWN

1

FIRE
HOSE
A/C   A/C

NORTH

CHAPTER 19

Banyan Sound
Elevator Foyer

OUTCOME SUMMARY

The intention of this exercise is to develop your proficiency to interpret and incorporate programmatic information into a design solution within a two-hour time limit. It is a Charette that will provide experience in working within a short deadline. The project simulates a request by your client to develop a preliminary solution after visiting the project site and meeting with the client.

client profile

Your client, W. William Neel, is rather delighted with the work you have already completed on his condominium in Long Island, and he would like to hear and see what your ideas would be for the elevator foyer of his beach condominium in Boca Raton. There are spectacular views of the intercoastal waterway and the ocean from inside the apartments. He owns apartment 26N and is sharing the expense of the renovation with his neighbor, Friedl and Edletraud Schmidt, who own apartment 26S. The Schmidts live in Germany and will be in Boca Raton for a few short days to review the project with Neel. He has asked you to meet with him for a few hours to review the project requirements, the existing conditions, and to establish a direction for the use and design of the space.

Neel's apartment is designed and furnished with traditional-style elements, whereas the Schmidts have created a classic, modern atmosphere in their apartment, including furnishings designed by Alvar Alto and Le Corbusier. The client realizes that these styles are completely different design approaches and he has agreed to a solution that incorporates a modern styling solution. Above all, they want it to be inviting and to capture a look in this foyer that will express "residence," rather than "elevator lobby," that cordially welcomes their visitors.

project details
General and Architectural Information

Location: Banyan Sound Condominiums, Elevator Foyer, Twenty-Sixth Floor, 550 South Ocean Boulevard, Boca Raton, Florida

Net usable square footage: 310

Project budget: $20,000.00. No bidding is required. General contractor has begun work.

Existing construction. Current facility plans and specifications are available. Site information gathering is required.

Codes and Governing Regulations

Occupancy: R-2, Residential

Building Type: Type I

Fire Suppression: Standpipe and Hose System

Design the project to comply with the most current edition of the SBCCI Standard Building Code (SBC), the National Electrical Code (NEC), the Uniform Fire Code (UFC), Florida Building Code, and the Americans with Disabilities Act.

EXISTING FINISHES

The foyer is partially renovated, including new 12″ × 12″ fossil stone tile on the floor and grass cloth wallcovering. The condominium entry doors are to be resurfaced with a burl wood veneer. The veneer has been applied to apartment 26S, and, as the photographs indicate, there is a problem with the door panel design telegraphing through the veneer. This door will be resurfaced with a new, better quality veneer and the appropriate backing material. A new crown molding has been applied to the wall at the ceiling level, and new low-voltage recessed lighting has been installed in the ceiling. These new finishes are to remain. Color images of the finishes can be found on the CD.

Figure 19–1
New Wood Finish,
Brookside Veneers

Figure 19–2
Wallcovering, Donghia

Figure 19–3
Flooring,
Alistair Mackintosh Ltd.

Planning
- Floor plan may include any furnishings that you find appropriate for the space.

Furniture
- Seating, tables, or console unit
- Wall sconces to replace existing ones
- Artwork
- Plants
- Accessories
- Area carpets

Design
- Select paint color for wood trim, ceiling, and elevator door fronts.
- Select new area carpets.

Equipment
- No equipment is required.

site visit findings

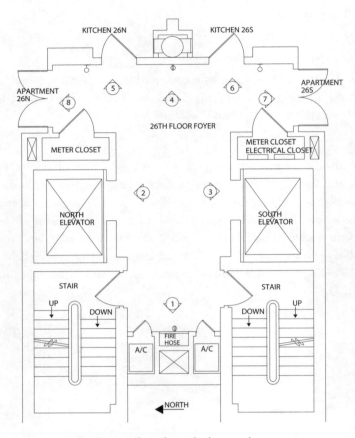

Figure 19–4   Floor plan with photograph notations

You have exited the north elevator, looked west (to your right) and see two doors that enclose the air conditioners. An access door is in the center of the wall enclosing a fire hose cabinet. There is a duplex electrical receptacle in this wall.

Figure 19–5   Photo 1: West Wall

The door to the left is the access door to the north fire stairs. The elevator door is located to the right.

Figure 19–6   Photo 2: North Wall

The door to the right is the access door to the south fire stairs. The elevator door is located to the left.

Figure 19–7   Photo 3: South Wall

The doors on this wall are secondary entrances to the condominium units. They provide access to and from the kitchens. There is a duplex electrical receptacle in this wall.

Figure 19–8   Photo 4: East Wall—Center

Figure 19–9   Photo 5: East Wall—Left

Figure 19–10   Photo 6: East Wall—Right

The door is the same secondary entrance door to the kitchen of Condominium Unit 26N as shown in Photo 4. The nameplate below the wall sconce is to remain. The wall sconce is to be replaced.

The door is the same secondary entrance door to the kitchen of Condominium Unit 26S as shown in Photo 4. The wall sconce is to be replaced.

This door is the entrance to Condominium Unit 26S. The substandard wood veneer finish is shown in this picture. The door on the right wall of the entrance door encloses an electrical closet. The door on the left wall is the door into the kitchen as shown in Photos 4 and 6.

Figure 19–11   Photo 7: South Wall—Apartment Entry Door 26S

This door is the entrance to Condominium Unit 26N, which is owned by the client. A new, higher quality, wood veneer finish by Brookside Veneers, will also be applied to this door. The door on the left wall of the entrance door encloses a closet. The door on the right wall is the door into the kitchen as shown in Photos 4 and 5.

Figure 19–12   Photo 8: North Wall—Apartment Entry Door 26N

SUBMITTAL REQUIREMENTS

Phase I: Programming

A. Read the background information, the client's program requirements, the existing conditions and limitations, and analyze the problem.

B. Formulate some visual ideas to express the character of the space. Use the scaled floor plan in Figure 19–13 and sketch your ideas on tissue overlays.

Evaluation Document: Programming Evaluation Form

Phase II: Conceptual Design

A. Develop a few simple tissue sketches of the plan before finalizing a floor plan.

B. Select appropriate furnishings and draw them to 1/4″ scale on the floor plan.

C. Select appropriate colors and materials to support your design solution.

D. Prepare a final floor plan on the plan provided as Figure 19–13. This plan may be drawn freehand or by using drafting equipment, but it must be drawn to 1/4″ scale.

E. Draw a freehand perspective sketch of the most important view of the foyer on tissue paper.

F. At the end of the two-hour time limit, present your solution to the class. Your verbal presentation must include justification for your spatial and functional solutions, as well as the visual character of the space designed.

Evaluation Document: Conceptual Design Evaluation Form

You will not need the CADD file for this exercise, but the AutoCAD LT 2002 drawing file name is banyan.dwg and can be found on the CD.

NOTES

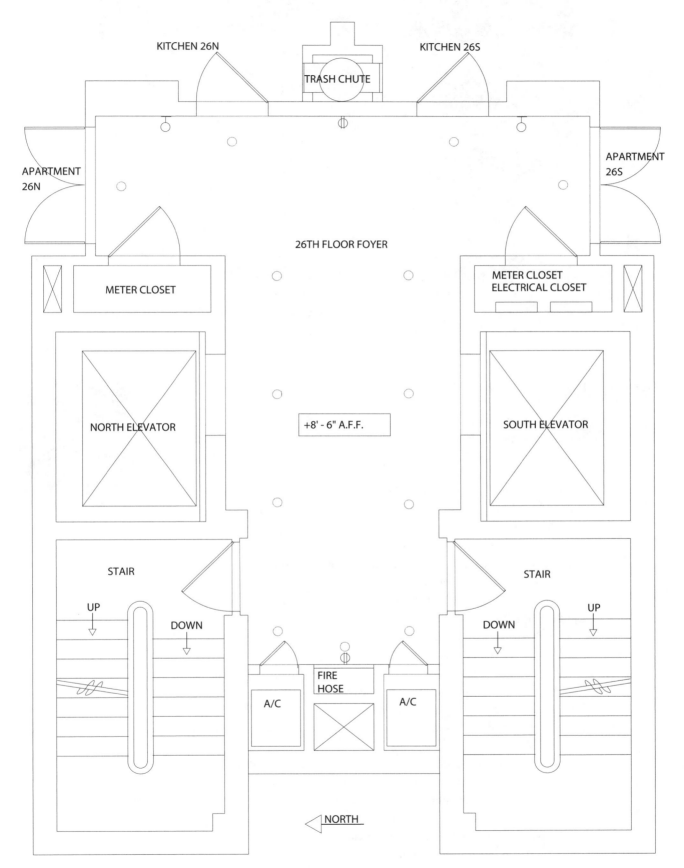

KITCHEN 26N

KITCHEN 26S

TRASH CHUTE

APARTMENT 26N

APARTMENT 26S

26TH FLOOR FOYER

METER CLOSET

METER CLOSET
ELECTRICAL CLOSET

NORTH ELEVATOR

+8' - 6" A.F.F.

SOUTH ELEVATOR

STAIR

STAIR

UP

DOWN

DOWN

UP

FIRE
HOSE

A/C

A/C

NORTH

Figure 19–13   Floor Plan. 1/4" = 1' 0" scale.

CHAPTER 20

IBACOS

OUTCOME SUMMARY

The project focuses on increasing awareness and responsibility of designers to minimize ecological and environmental impacts and the related programs in current building practice. Also included is research of how today's technology is available for an intelligent home. The project features three phases of the design process: identifying and analyzing the client's goals, developing conceptual skills through schematic or initial design concepts, and detailing and refining ideas from the schematic design phase.

client profile

Figure 20–1
Exterior view of model home. *(All photographs Courtesy of IBACOS, Inc.)*

You are to design eco-friendly, visual details, and architectural features and elements to enhance specific wall planes of the interior space of the Entry, Home Office, Master Bedroom, and Cupola while paying special attention to the principles of a *Better Home* as outlined below. For example, what would be the alternative to wood panels or moldings on a wall? What would be a more responsive and efficient use of natural resources? The wall planes have been finished with drywall and defined in the following figures in plan and elevation.

The project is a demonstration home to aid in the education of builders, trades, and homeowners sponsored by Integrated Building and Construction Solutions (IBACOS). This pilot home will be used to test and validate performance and cost results of systems integration in the field and was built by Montgomery & Rust. The home is located in the mid-Atlantic region of Northeastern United States and "features

- High performance low-E windows
- Increased insulation values
- Complete air sealing package to minimize energy loss through the thermal envelope
- High-efficiency furnace, air conditioner, and water heater
- Minimized duct losses by centralized furnace
- Perimeter and under-slab insulation reduce heat loss to the ground and provide moisture control
- Low-water consumption toilets, clothes washer, and dishwasher
- Programmable thermostat and multi-zone mechanical dampering control system
- Open floor trusses minimize job site waste and productivity
- Engineered wood studs, beams and window/door headers provide efficient use of trees by using less wood per size and strength material
- Compact fluorescent fixtures provide quality lighting with significant energy savings.[1]

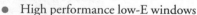

1. U.S. Department of Energy.

188

In neighborhoods throughout the country IBACOS's goal has been realized through the building of a new generation of homes that embody their concept of value and livability. Their concept of "Resource Responsibility," that is, reducing the ecological and environmental impacts by giving homes design consideration and optimization at the inception of community projects, has gained both recognition and support from both public and private sectors. According to IBACOS, "Value and livability are maximized around the following defined principles of a *Better Home*:

- For individual lifestyle fit,
- Ease of use and maintenance,
- Occupant safety and comfort,
- Quality construction, durability and craftsmanship,
- Responsible and efficient resource use,
- In connection with the surroundings and
- Affordable home ownership and operation."

IBACOS works in the company of some of the nation's leading manufacturers, builders, developers, and experts in the home-building industry with a common interest in creating and delivering a *Better Home*. Manufacturer partners are: USG Corporation, manufacturer of gypsum panels, joint compound, and related building products; Carrier Corporation, manufacturer of HVAC equipment and system controls; Owens Corning, producer of glass fiber reinforcements, textile yarns, and insulation; Andersen Corporation, manufacturer of high-performance windows and doors; and Burt Hill Kosar Rittelmann, experts in architectural, engineering, and energy systems design.

Government partners of IBACOS are: the Department of Energy's Building America Program; the National Renewable Energy Labs; Oak Ridge National Laboratory; Lawrence Berkeley National Labs; EPA Energy Star® Program; and Partnership for Advancing Technology in Housing[2] (PATH).

project details

General and Architectural Information

Location: Washington's Landing, 40 Waterfront Drive

Gross square footage: 3,920. Four Levels: Basement, First Floor, Second Floor, and Attic

New construction. Architectural plans are available. The architect/contractor will administer construction documents and the bidding of the interior furniture, fixture, and finish work.

Consultants required: Architect, electrical engineer, and mechanical engineer

Codes and Governing Regulations

Design the project to comply with the International Residential Code (IRC).

Occupancy: R-4. Residential, Individual One-Family Dwelling

The project was originally designed to comply with the BOCA National Building Code (NBC), the National Electrical Code (NEC), the Uniform Fire Code (UFC), the Americans with Disabilities Act, and the current editions of local community codes.

Facilities Required

Entry, Home Office, Master Bedroom, and Cupola

---

2. Courtesy IBACOS, Inc.

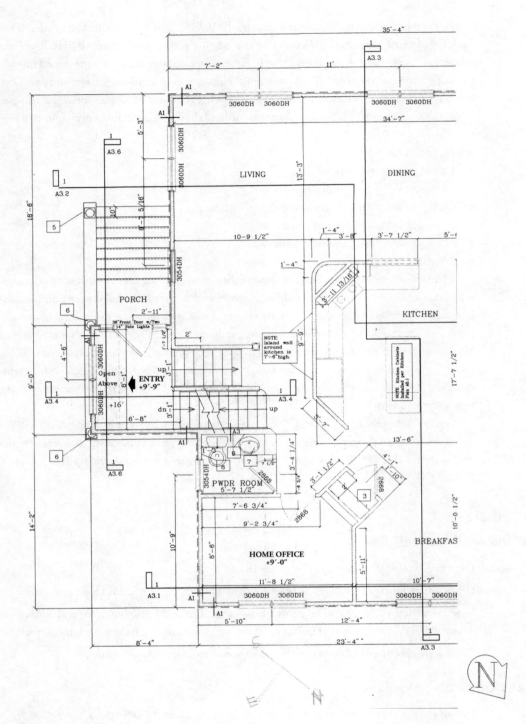

Figure 20–2
Floor Plan, First Level. *(All floor plans and sections courtesy of IBACOS, Inc.)*

ENTRY

● The split entry is a multi-level experience with distinctive architectural features, including a dramatic three-story window wall, large expanses of wall surface, and a traditional stair configuration with a balcony on the second level.

● Concentrate your design on the large window wall (Figure 20–3) while keeping in mind that it is visible from three levels.

HOME OFFICE

● This is the nerve center of the home and includes the operational panels for the mechanical and electrical systems. It will be used as a typical home office with complete computer system integration. Limit your design to the wall surfaces of this room.

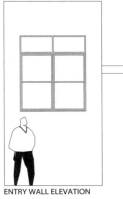

ENTRY WALL ELEVATION

Figure 20–3
Elevation of Entry Wall

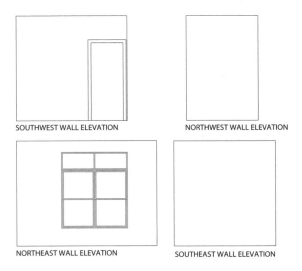

SOUTHWEST WALL ELEVATION   NORTHWEST WALL ELEVATION

NORTHEAST WALL ELEVATION   SOUTHEAST WALL ELEVATION

Figure 20–4
Elevations of Office Walls

MASTER BEDROOM

- The bedroom has an open plan concept and includes the sleeping and grooming areas.
- Direct your focus to the wall separating the sleeping area from the grooming area, as shown in Figure 20–5. This is the wall plane you are to design. Wall elevation can be found on the CD, in the drawing file ibacos.dwg.

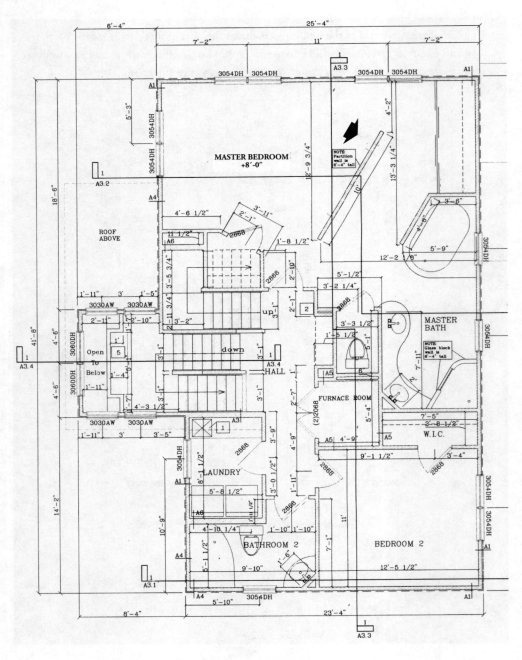

Figure 20–5
Floor Plan, Second Floor. *(Courtesy of IBACOS, Inc.)*

CUPOLA

- The cupola is a dramatic architectural element that adds additional light and depth to the space. It is located on the attic level in a room that may be dedicated to a number of activities, depending on the final owners of the home.
- Solve the issue of providing lighting in the cupola space during a cloudy day or the evening hours.

Figure 20–6   Installation of Cupola.

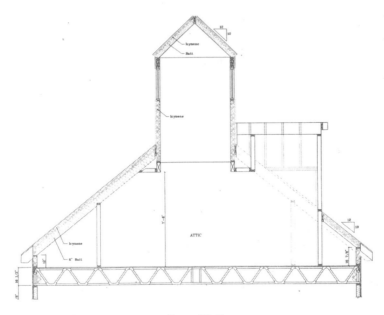

Figure 20–7
Section Elevation of Cupola

SUBMITTAL REQUIREMENTS

Phase I and Phase II: Programming and Conceptual Design

Part A Due: _____   Part B Due: _____   Part C Due: _____

Part D Due: _____

A. Select a manufacturer from the list to feature its products in one of your solutions.

B. Research ecological and environmental impacts and the related programs in current building practice and responsible and efficient resource use in interior design practice today. Research how today's technology can be incorporated into a "smart home."[3]

C. Hold a design team meeting dialogue group for the project. In the meeting discuss the topic of resource responsibility and propose strategies to achieve this for the interiors. In the dialogue group session,, share your knowledge from the research you have completed and arrive at a design approach for the project.

D. Use the golden ratio (golden section)[4] proportioning system to create the visual relationship of the design element proportions for one of the wall elevation concepts.

Evaluation Documents: Programming Evaluation Form and Conceptual Evaluation Form

Phase III: Design Development

Part A Due: _____   Part B Due: _____   Part C Due: _____

A. Prepare a plan, indicating any special design elements and attached interior fixtures keyed to details, sections, and elevations as required.

B. Prepare elevations or details of walls, showing design elements, fixtures, and color and material indications keyed to details and samples.

C. Details showing sections through specially designed elements or fixtures must be prepared and identified as to wall locations.

Evaluation Document: Design Development Evaluation Form

Phase IV: Final Design Presentation   Date Due: _____

A. The final presentation will be an oral presentation to a panel jury. Project must be submitted on 20″ × 30″ illustration boards. Your name, the date, the project name and number, the instructor's name, and the course title must be printed on the back of each board in the lower left-hand corner. Submit as many boards as required, including plans, elevations, and material selections.

B. The plans should be presented in 1/4″ scale. The elevations should be presented in 1/2″ scale. The plans and elevations are to be completed using the CADD program. Print drawings on 11″ × 17″ bond paper.

Evaluation Document: Final Design Presentation Evaluation Form

Project Time Management Schedule: Schedule of Activities, Chapter 31.

The AutoCAD LT 2002 drawing file name is ibacos.dwg and can be found on the CD.

---

3. Lee, Jenne, Brian L. Clar, Art Janik, and Amy Wilson. 2002. Smart homes: Gadget gallery. MONEY Magazine, September.
4. *Designing Interiors. Interior Design Illustrated. Interior Design, 2nd ed.*

REFERENCES

Books

Ching, Francis D. K. 1987. *Interior design illustrated.* New York: Wiley.

Kilmer, Rosemary, and W. Otie. 1992. *Designing interiors.* Fort Worth, Tex.: Harcourt Brace Jovanovich College Publishers.

Pile, John F. 1995. *Interior design.* 2d ed. New York: Abrams.

Special Resources

Office of Building Technology, State and Community Programs, U.S. Department of Energy, Building America Program: http://www.eren.doe.gov/buildings/

IBACOS, Inc. (Integrated Building And Construction Solutions) http://www.ibacos.com

Internet Home Alliance: http://www.internethomealliance.com

Product Manufacturers

Anzea: http://www.anzea.com

Carrier Corporation: http://www.carrier.com

GE Corporation: http://www.ge.com

Home Automated Living & HAL: http://www.automatedliving.com

Kohler Company: http://www.kohler.com

Owens Corning: http://www.owenscorning.com

USG Corporation: http://www.usg.com

CHAPTER 21

Neel Condominium Residence

OUTCOME SUMMARY

The project features four phases of the design process: identifying and analyzing the client's needs and goals, developing conceptual skills through schematic or initial design concepts, detailing and refining ideas from the schematic design phase, and drafting documents in preparation for the bidding and contracting of construction, fixtures, and furnishings. The project focuses on increasing knowledge of design practice for high-rise, urban condominium residences and the design related rules and regulations.

client profile

William Neel is a classy and fun-loving client who has a great affection for art, sports, and travel. He is part owner of one of the city's major league sport teams and president of a general contracting, construction, and construction management firm. The company builds commercial, institutional, power, industrial, bridge, and highway projects in North America, Europe, Latin America, and the Middle East. An active community member, Neel, in his early forties, is practical and fun loving. You may find him spending his free time playing golf, tennis, and racquetball, and working out. He enjoys art, music, and reading, and travels frequently for business and pleasure. Neel has an extensive family art collection and will be using many of the works in this home.

Neel has purchased a city condominium to reside in until his home renovations are complete and has employed you to renovate the space. Because his home, cottage, and beach residences are designed in a formal traditional style, he would prefer that the urban condo be more casual. A clear and honest expression of an appropriate interior solution is to be reflected in your recommendations to maintain the design consistency and integrity established for the project. The Grand is a prestigious high-rise condominium on Mountview Avenue situated on the hills overlooking the downtown skyline with a commanding view of the urban scene, by far its greatest asset. It is also uniquely positioned for watching the frequent firework displays hosted by the city numerous times a year. Along the avenue there are several lookout points, upscale dining facilities, and two historic cable car inclines that are still in service.

The building features include hospitality suites for guests, a business center and conference room, a party and game room with catering kitchen and championship pool table, a private fitness center, climate-controlled storage areas, a resort-style pool and whirlpool spa, a grill area, and 24-hour security.

The condominium association regulations particular to your scope of work are as follows:

- Nothing shall be done in any unit or on the common areas which may impair the structural integrity of the building or which may structurally change the building. Nothing may be removed from or changed in the common areas without the prior written consent of the Board of Directors. (This applies to the remodel of the lobby area.)

- All structural repairs and replacements shall be of first-class quality and as similar to the character of the construction or installation that existed prior to the occasion that necessitated the repairs or replacements. Repairs and replacements shall be done with contemporary building materials and equipment after all necessary permits, certificates of insurance, and approvals have been obtained.

- All radio, television, sound system, or other electrical equipment of any kind or nature installed or used in each unit shall fully comply with all rules, regulations, requirements or recommendations of the Board of Fire Underwriters and the public authorities having jurisdiction and the unit Owner alone will be liable for any damage or injury caused by any radio, television, sound system, or other electrical equipment in each unit.

- To ensure a uniform, attractive external appearance to the building, all window treatments must show an exterior (that is, the side facing outward) color that is either white or off-white. (If other colors are used, they must be backed with a white or off-white lining so that the lining shows to the outside rather than the colored or decorative side.) Sheets, plastic, cardboard, plywood and other such materials are prohibited except for limited temporary use following a casualty to a unit.

- All floors in each unit shall have a proper sound-absorbing underlayment to adequately limit sound transmission to other units. The board of directors must approve installation of any floor covering, except carpeting in order to ensure proper sound-absorbing methods are being adhered to.

- To ensure structural integrity and sanitary conditions, the following rules apply:
    1. Nothing will be affixed to the exterior building facing or ceiling of a unit balcony area.
    2. To prevent water accumulation or damage, balcony floors will not be covered with carpet.
    3. The attachment of satellite dishes to balconies or railings is prohibited.
    4. No balcony shall be enclosed or covered and there shall be no awnings, canopies, blinds, shades, screens, or similar fixtures attached to, hung in, or used in connection with any balcony of a unit.

- Customary lawn or patio furniture in good condition and reasonably sized, and well-tended plants may be placed on the balcony. However, all items should be placed and secured so as to protect against being blown or pushed off the balcony.

- In order to ensure a uniform appearance from the outside of the building, no unit shall hang, display, or expose, from any window, door, balcony, or exterior of the unit, so as to be visible from anywhere on the Grand property, any of the following:
    1. Signs, posters, or decorations, with the following exceptions: the American flag (appropriately displayed) and temporary appropriate seasonal decorations.

- No charcoal cooker, brazier, hibachi, grill, or any gasoline, propane, or other flammable liquid, or liquefied petroleum gas-fired stove, or similar devices shall be ignited or used on balconies or within the units under order of the county fire marshal.

- Deliveries of furniture, appliances, and so forth, and move-ins/move-outs, must use the loading dock. Reservations to use the elevator and loading dock must be scheduled in advance, at the management office, between 8:00 a.m. and 4:00 p.m., Monday through Friday.

- Delivery of any items too large for the interior elevator cab must be scheduled in advance at the management office. A payment of $250.00 is required for the elevator company to transport the item on top of the elevator cab.
- Management can refuse deliveries not scheduled in advance. If the resident will not be home, prior arrangements should be made at the front desk for access to the unit by means of the resident providing both an admit slip and key for the specified delivery.

project details

General and Architectural Information

Location: The Grand, Unit 2400 A

Gross square footage: 1,930

Project budget: $120,000.00 furniture budget. The architect/contractor will administer construction documents and the bidding of the interior furniture, fixture, and finish work.

Existing construction. Ceilings and floors are concrete and the walls are drywall. The maximum available ceiling height of 9 feet occurs in each space and drops to 8 feet in the corridors, where it hides the mechanical systems. Architectural plans are available. Start construction in February.

Consultants required: Architect, electrical engineer, and mechanical engineer

Codes and Governing Regulations

Occupancy: R-2, Residential

Building Type: Type I

Fire Suppression: Standpipe and Hose System

Design the project to comply with the most current edition of the BOCA National Building Code (NBC), the National Electrical Code (NEC), the Uniform Fire Code (UFC), and the Americans with Disabilities Act. The minimum habitable room size is 120 square feet. The minimum ceiling height for habitable rooms is 7′ 0″. The minimum handrail height is 34″. Smoke alarms must be installed in every bedroom, any hallway leading to a bedroom, and on each level of a multistory building. Safety bars on windows must release easily in an emergency. Windows and hallways should be clear of obstructions. Use great judgment when placing large pieces of furniture in front of windows that are an escape route.

Facilities Required

Elevator Vestibule, Entry, Living Room, Dining Room, Guest Bathroom, Kitchen, Balcony, Master Bedroom, Master Bath, Master Closet, Guest Room/Office, and Laundry

ELEVATOR VESTIBULE

Furniture

- Plants; accessories, including an umbrella stand; and furnishings that you find appropriate for the space

Design

- Create a warm and inviting notion of arrival in the small lobby area. As a guest it is not simple to reach the front door: parking is a block away, the windy walk to the building is uphill. Upon entering the lobby a sign-in and call-up procedure begins with the security attendant, and, finally, the elevator is the last leg of the experience.

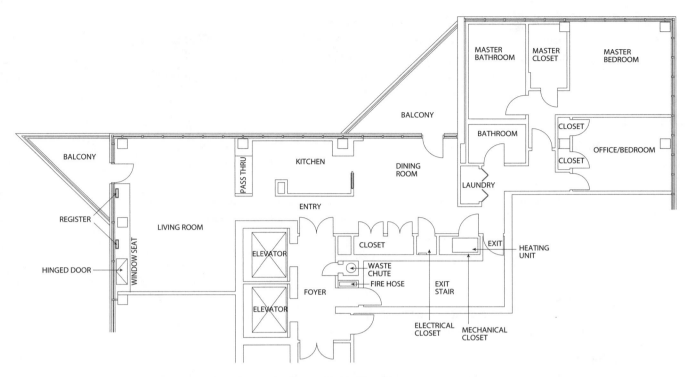

Figure 21–1   Floor Plan

- The board has given approval to make the following alterations to the lobby: new ceiling design and finish.
- Provide indirect lighting for the space and feature lighting for the artwork.
- Provide new paneled doors, wallcovering, paint, trim and molding, and hard surface flooring with an area carpet.

Equipment

- A new lockout elevator key will be installed.
- Fire alarm and fire hose cabinet are to remain.

ENTRY

Planning

- This space is at the lower ceiling height and opens laterally to a living area on the left side and the dining area on the right side.
- The entry area is the only access to the home and leads to the Living Room and Dining Room.
- Develop a furniture plan that directs circulation and provides convenience for Neel.
- Provide coat and hat storage in one section of the double door closet wall.

Furniture

- Any furnishings that you find appropriate for the space. Provide a place to drop keys, mail, etc.
- Plants and accessories
- Area carpet

Design

- Provide a hard surface floor material. Carefully consider its transition into the Living and Dining Rooms.
- Provide selection of wall and ceiling finishes.

- Provide custom detailing for base, trim, and new panel doors.
- Incorporate a lighting solution for the Entry that will provide general illumination and feature artwork. This same lighting is to be used in the dining area.
- Study all design transition elements and finishes as they relate to adjacent rooms and the hallway to the private zones of the home. Use logical solutions to material changes.

LIVING ROOM

Planning

- The Living Room is situated at the front of the home to take advantage of the unobstructed vista.
- This room is used primarily to watch television, listen to music, read or relax, and entertain guests and should be an atmosphere favorable to conversation. Friends and guests visit frequently.
- Provide a comfortable and intimate seating arrangement to accommodate approximately six to eight people.

Furniture

- Sofas, chairs, counter stools, tables, lamps, accessories, area carpets
- Two bar stools at the counter area

Design

- Design a custom wall unit with equipment hidden behind cabinet doors so it is integrated into the overall design. Provide lighting in the wall unit. Provide storage for tapes, DVDs and CDs in pullout drawers.
- Provide carpet floor covering.
- Provide selection of wall and ceiling finishes.
- Provide custom detailing for bases and trims.
- Window treatments are required. There is a fantastic view of the skyline from this room but intense sunlight in the afternoon hours.
- Provide window seat cushions and plenty of back pillows between the wall and column along the window. The seat pillows must be removable to access the door enclosing the heat pump. Neel has asked for a ball-shaped "throw" pillow about the size of a soccer ball.

Equipment

- Complete home entertainment system, including television, VCR and DVD equipment, and stereo sound system
- Telephone

DINING ROOM

Planning

- The dining room is used extensively for entertaining guests both formally and informally.
- Provide a round table that will seat six people. Larger parties will be served buffet style.
- Storage is required for china, other tableware, silver, glassware, and linens.
- Plan for wine storage in one section of the double door closet wall and include wine cooling system.

Furniture

- Dining table
- Dining chairs for six people
- Buffet and server

- Lamps, accessories, area carpets, and any other furnishings that you find appropriate for the space

Design

- Provide selection of wall and ceiling finishes.
- Provide custom detailing for bases, trims, and new panel doors.
- Incorporate a lighting solution that will provide general illumination and feature artwork. This same lighting is to be used in the Entry. Incorporate a pendant or chandelier arrangement over the dining table.
- Window treatments are optional and must not interfere with access to the Balcony.

GUEST BATHROOM

Planning

- Both guests and family members will use the bath.
- Provide storage for linens.

Furniture

- Bath accessories, large towel hooks, door hook, mirrors, and area carpets

Design

- Provide hard surface floor covering.
- Provide selection of wall and ceiling finishes.
- Provide custom detailing for bases, trims, and new panel doors.
- Provide specialty fixtures and casework.

Equipment

- Lavatory, vanity, water closet, and bathtub/shower combination

KITCHEN

Planning

- The room size and shape calls for a "corridor"-type kitchen layout. Develop a functional and convenient arrangement around a "work triangle."[1]
- Provide a triple stainless steel gourmet sink system with dispensers and two-drawer dishwasher area with counter space on each side. Food disposal must be located in center sink.
- Provide Sub Zero refrigerator/freezer and adjacent counter area.
- Provide a cooking area with counter space on each side. Cooktop should include modular interchangeable elements for the grill, griddle, wok, and rotisserie. Provide proper ventilation.
- Neel has requested that the counter heights be at 38″ because of his tall stature. Wall cabinets must meet ceiling trim at the top.
- Provide wall and base cabinets for storage of linens, food, cookbooks, utensils, bowls, storage ware, small appliances, serving ware, dishes, flatware, glassware, cookware, and bake ware.
- Special cabinet features, including recycle waste containers, under sink pullout towel bars, divided cutlery and flatware drawers, slanted spice rack holders in drawers, upright plate slots, pullout drawer units at all base cabinets, extra pullout work surface made of granite surface material, tiltout storage at sink cabinet, and deep and tall drawers for cookware are required.

---

1. Designing Interiors. p. 224.

- Provide base cabinet and granite countertop (height to be 38″) at area between the Kitchen and Living Room, which is an oversized pass-through. This area will be used for casual dining, buffet serving, or bar setup area while entertaining.

Design

- Provide selection of wall and ceiling finishes.
- Provide hard surface flooring.
- Granite countertops and backsplash are required. Incorporate honed drain area in the countertop adjacent to the sink. Provide granite shelves above the sink area for orchid plants.
- Provide custom detailing for bases, trims, and new panel doors.
- Provide window treatment.

Equipment

- Refrigerator, dishwasher, cooktop, hood, oven, microwave, and trash compactor
- Triple sink with disposal
- Telephone
- Fire extinguisher

BALCONY

Planning

- Living Room Balcony: The winds are very strong on this side of the building and the balcony will be used during entertaining if guests would like to smoke a cigar or enjoy the city panorama.
- Plan for high bar chairs and side tables and low plants on the Living Room Balcony.
- Dining Room Balcony: Plan a seating arrangement for conversation and dining and a lounging group.

Furniture

- Settees, chairs, ottomans, chaise lounges, tables, and planting pots

Design

- Provide new tile floor covering and base trim.
- All materials must be suitable for outdoor use.

MASTER BEDROOM

Planning

- It is primarily a sleeping area but will also function as a private retreat for relaxing, watching television, or reading.

Furniture

- King-size bed with a deep mattress and related furniture
- Personal storage
- Comfortable lounging chair and related furniture
- Lamps, accessories, and any other furnishings that you find appropriate for the space

Equipment

- Television
- Telephone

Design

- Provide custom bedding.
- Provide carpet floor covering.
- Provide selection of wall and ceiling finishes.

- Provide custom detailing for bases, trims, and new panel doors.
- Window treatments with blackout capability are required.

MASTER BATH

Planning

- Provide storage space for items related to grooming, bath items, and linens.

Furniture

- Magnifying mirror attached to an accordion arm located above the lavatory counter and in the shower
- Bath accessories, door hooks, mirrors, and any furnishings that you find appropriate for the space

Design

- Provide granite floor and walls, tub, and shower. Shower enclosure must have clear glass. Design a seat in the shower.
- Provide selection of wall and ceiling finishes.
- Provide ample task lighting for shaving.
- Provide custom detailing for cabinets, bases, trims, and new panel doors.
- 38″ counter height is required.
- Specialty fixtures and casework

Equipment

- Two lavatories, water closet, whirlpool bath, and double showerhead with additional wall water jets, with the main showerhead positioned high enough
- Telephone

MASTER CLOSET

Planning

- This dressing area requires storage for hanging and folded clothing, shoes, and accessories.
- Incorporate a built-in ironing center.

Design

- Relate to the bedroom scheme.

GUEST ROOM/OFFICE

Planning

- This guest room, a dual-function space, is used primarily as a home office and occasionally for overnight guests.
- Design a custom wall unit to function as an office. Provide a desk area, storage for work-related items, file storage, book storage, entertainment closed-door function, and storage for folded clothing.
- Provide hanging and shelving in the two built-in wall closets.

Furniture

- Sleep sofa, ottoman, and related furnishings
- Lamps and accessories

Design

- Provide carpet floor covering.
- Provide selection of wall and ceiling finishes.
- Provide custom detailing for bases, trims, and new panel doors.
- Custom bedding and window treatment with blackout capability are required.

Equipment

- Laptop computer with docking station, printer, fax, copier
- Telephone
- Television and VCR/DVD

LAUNDRY

Planning

- Plan for convenience and function, including the necessary appliances and fixtures.

Design

- Provide selection of wall and ceiling finishes.
- Shelving and/or cabinet storage, bases, trims, and new panel doors are required.
- Provide hard surface flooring.

Equipment

- Washing machine, clothes dryer, and deep laundry sink

SUBMITTAL REQUIREMENTS

Phase II: Conceptual Design   Date Due: _____

A. Develop a preliminary plan based on the functional and aesthetic requirements necessary, including use of space, furniture layouts, and fixtures.

Evaluation Document: Conceptual Design Evaluation Form

Phase III: Design Development   Date Due: _____

A. Based on the approved preliminary plan, develop and present for review and approval, final interior design recommendations to fix and describe the character of the interior space of the project.

B. Prepare documents and illustrations in the form of plans, elevations, four to six color and material sample boards, photographs of furnishings and fixtures, and two color renderings to show:

    a. Wall, furniture, and fixture arrangement
    b. Floor, wall, window, and ceiling treatments
    c. Furnishings, fixtures, and millwork
    d. Schematic lighting plan and lighting fixtures
    e. Colors, materials, and finishes
    f. Suggestions for art, accessories, and plants

Evaluation Document: Design Development Evaluation Form

Phase IV: Final Design Presentation   Date Due: _____

A. The final presentation will be an oral presentation to a panel jury. Project must be submitted on 20″ × 30″ illustration boards. Your name, the date, the project name and number, the instructor's name, and the course title must be printed on the back of each board in the lower left-hand corner. Submit as many boards as required, including renderings, elevations, fixture and furniture selections, material selections, and plans.

    B. The plans should be presented in 1/4″ scale. The elevations should be presented in 1/2″ scale.

Evaluation Document: Final Design Presentation Evaluation Form

Phase V: Construction Documentation   Date Due: _____

    A. Prepare working drawings and specifications for non-load-bearing interior construction, materials, finishes, furnishings, fixtures, and equipment for the client's approval. The drawings and specifications must include:

        1. Plan indicating any special design elements and attached interior fixtures keyed to details, sections, and elevations as required.

        2. Elevations or details of walls showing design elements, fixtures, and color and material indications keyed to details and finish specification schedules.

        3. A floor covering plan and details keyed to specification schedules.

        4. Details showing sections through specially designed elements or fixtures must be prepared and identified as to wall locations.

        5. A lighting criteria plan must be provided and keyed to a lighting fixture schedule.

        6. An electrical power criteria plan must be provided and keyed to a legend. Plans or elevations are to include stereo speakers and smoke detectors.

        7. A reflected ceiling plan indicating new ceiling conditions, materials, and details with specifications must be prepared. Plans or elevations are to include stereo speakers and smoke detectors.

        8. All materials and finishes must be specified on color and material schedules.

    B. The plans should be presented in 1/4″ scale. The elevations should be presented in 1/2″ scale.

Evaluation Document: Construction Documentation Evaluation Form

Project Time Management Schedule: Schedule of Activities, Chapter 31.

The AutoCAD LT 2002 drawing file name is neel.dwg and can be found on the CD.

REFERENCES

Books

Ching, Francis D. K. 1987. *Interior design illustrated.* New York: Wiley.

Clodagh. 2001. *Total design, contemplate, cleanse, clarify, and create your personal spaces.* New York: Clarkson Potter.

DiChiara, Joseph, Julius Panero, and Martin Zelnik. June 2001. *Time-saver standards for interior design and space planning.* 2d ed. New York: McGraw-Hill.

Harmon, Sharon Koomen, and Katherine E. Kennon. 2001. *The codes guidebook for interiors.* New York: Wiley.

Kilmer, Rosemary, and W. Otie.1992). *Designing interiors.* Fort Worth, Tex.: Harcourt Brace Jovanovich College Publishers.

Nissen, LuAnn, Ray Faulkner, and Sarah Faulkner. 1994. *Inside today's home.* 6th ed. Fort Worth, Tex.: Harcourt Brace Jovanovich College Publishers.

Panero, Julius. 1989. *Human dimension and interior space: A source book of design reference standards.* Norwell, Mass.: Kluwer Academic.

Pile, John F. 1995. *Interior design.* 2d ed. New York: Abrams.

Trade Magazines

Interior Design Buyers Guide: http://www.interiordesign.net

Product Manufacturers

Elkay Manufacturing Company: http://www.elkay.com
Miele Appliances: http://www.mieleusa.com
Moen, Inc.: http://www.moen,com
Vinotemp International: http://www.vinotemp.com
Vola: http://www.vola.dk
Boyd Lighting Company: http://www.boydlighting.com
Lutron Electronics Co., Inc.: http://www.lutron.com
Tech Lighting LLC: http://www.techlighting.com
Brookside Veneers, Ltd.: http://www.veneers.com
Brunschwig & Fils: http://www.bruinschwig.com
Decortex/Firenze: http://www.decortex.com
Hunter Douglas: http://www.hunterdouglas.com
Kravet: http://www.kravet.com
Maya Romanoff Wallcovering: http://www.mayaromanoff.com
Scalamandre: http://www.scalamandre.com
Kohler: http://www.kohler.com (Kohler, Baker, Anne Sacks)
Walker& Zanger, Inc.: http://www.walkerzanger.com

NOTES

CHAPTER 22

Hahn Residence

OUTCOME SUMMARY

The project features four phases of the design process: identifying and analyzing the client's needs and goals, developing conceptual skills through schematic or initial design concepts, detailing and refining ideas from the schematic design phase, and drafting documents in preparation for the bidding and contracting of construction, fixtures, and furnishings. The project focuses on increasing knowledge of design practice for single-family residences and the development of a future accessible apartment for the elderly.

client profile

Charles, Mitzi, and Marissa Hahn are ready to build their first home on two acres of rural land, a 20-minute drive from the city. The single-story casual hideaway structure has a full basement. It will be perched at the top of a hill and positioned to take advantage of the wooded vistas. Views are important in the Living Room, Dining Room, and Kitchen. The numerous windows in their unique geometrical pattern will flood the interior with light and create a sense of airiness. They want the home to be somewhat transparent from front to back and reveal a large sunny space inside. A hindrance will be the intense sunlight, so they plan to install a retractable awning on the deck side of the house. You will have to give special consideration to a nonobtrusive window covering for these window walls at the front and back.

The home's exterior will be clad with wood siding. The natural materials of the exterior introduce a feeling that blends well with the woody surroundings. Having a great affinity to modern design and crisp contemporary palettes, they want you to gradually blend the exterior with a modern interior while making it innovative in design and materials. Intriguing spaces and distinctive details are synonymous with their personalities. And rest assured that if you come up with a distinctive detail that stumps you in terms of construction, Charles will be determined

Figure 22–1   Rear Exterior Elevation

Figure 22–2   Front Exterior Elevation

to find the solution and make it happen. They prefer light colors, such as white and silvery greens, and clean-looking materials, such as glass, satiny metals, and sleek lacquered surfaces.

A working couple in their mid-thirties with one 2-year-old daughter, they live at a very accelerated pace that includes a 24/7 telephone life. Mitzi lives a healthy lifestyle, which includes exercise and good nutrition. She owns and manages three trendsetting women's clothing shops, loves her work, entertaining, relaxing, and, most importantly, spending time with her gregarious daughter Marissa. Charles is equally busy with several business ventures: one as a partner in a company that sells mechanical equipment, and the other as the owner of a Harley Davidson dealership in Las Vegas. He is stylish, decisive, knows what he wants, and has an especially strong personality. When he is not riding his custom Soft Tail Harley Davidson, he might be off traveling the Indy circuit with his buddy who owns a race car team.

As a family the Hahns enjoy boating and skiing and any type of water sport. Their immediate families visit frequently, including all four parents and the grown brothers and sisters. All encourage every opportunity for affection and togetherness.

A sense of calm and retreat is the most important aspect to consider in the home. The Hahns want a flexible living environment for informal gatherings and casual entertaining. In order to do this well in a small amount of space, the living area, dining area, and kitchen is to be one symmetrical, open plan space. To maximize the square footage, the plan is to be orderly and the design informal. Charles prefers to use track lighting because of its flexibility of use.

They anticipate a need for an accessible apartment on the lower level for an elderly parent with limited mobility. You are to design an alternative plan for the lower level that includes a living area, dining area, kitchenette, bedroom, and bathroom that are completely accessible. You are also to study a location to place an elevator in future years.

Provide smoke and carbon monoxide detectors throughout the home.

project details

General and Architectural Information

Gross square footage: Two Stories: 4,200. First Level: 2,080 square feet. Lower Level: 2,120 square feet

Project budget: Mid range. The architect/contractor will administer construction documents and the bidding of the interior furniture, fixture, and finish work.

New construction. Architectural plans are available.

Consultants required: Architect, electrical engineer, and mechanical engineer

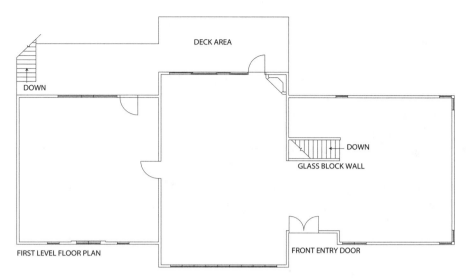

Figure 22-3 First Level Floor Plan

Codes and Governing Regulations

Occupancy: Residential, Individual One-Family Dwelling

Design the project to comply with the most current edition of the International Residential Code (IRC) and the Americans with Disabilities Act. The minimum habitable room size is 120 square feet. The minimum ceiling height for habitable rooms is 7′ 0″. The minimum handrail height is 34″. Smoke alarms must be installed in every bedroom, any hallway leading to a bedroom, and each level of a multistory building. Safety bars on windows must release easily in an emergency. Windows and hallways should be clear of obstructions. Use great judgment when placing large pieces of furniture in front of windows that are an escape route.

Facilities Required

First Level: Entry, Living Room, Dining Room, Kitchen, Bathroom, Marissa's Room, Guest Bedroom, Master Bedroom, Master Bath, Master Closets, Laundry, and Deck

Lower Level: Family Room, Office, Guest Bathroom, Wine Cellar/Storage, and Pool Patio. (Alternate Plan for Lower Level: Future accessible apartment)

First Level Spatial Requirements

ENTRY

Planning

- Adjacency: Front entry door and Living Room
- The entry is the main entrance for greeting guests and is the transition space into the Living Room, private zones, and lower level of the home. This will be a "shoes off" household.
- Develop a furniture plan that directs circulation as well as making a first impression upon entering.
- Provide storage for coats, hats, boots, and umbrellas.

Furniture

- Console table, a place to drop keys, etc., and any furnishings appropriate for the space.
- Plants, accessories, and artwork
- Area carpet

Design

- Provide a hard surface floor material. Carefully consider its transition into the Living and Dining Rooms.
- They have requested a glass block feature wall to enclose the stair passage. The stair configuration may be redesigned.
- Provide selection of wall and ceiling finishes.
- Provide custom detailing for bases and trims.
- Incorporate a lighting solution for the entry that will provide general illumination and feature artwork. Provide dimmers for the general illumination.
- Study all design transition elements and finishes as they relate to adjacent rooms and the passage to the private zones of the home. Use logical solutions to material changes.

LIVING ROOM

Planning

- Adjacency: Entry, Dining Room, and Deck
- This room is used primarily to watch television, listen to music, read or relax, and entertain guests, and should be an atmosphere favorable to conversation. They will often gather around with friends, family, and guests over a glass of wine and cocktail snack.
- Provide a comfortable and intimate seating arrangement to accommodate approximately ten to twelve people.
- Incorporate a complete home entertainment system, including television with a 30″ screen, VCR and DVD equipment, and stereo sound system. This equipment is to be housed in a contemporary, custom wall unit with glass doors. Provide task lighting for the wall unit. Charles has an extensive music collection and needs six to eight linear feet of storage for tapes, DVDs and CDs.
- This space must be an open plan and have direct access to the outdoor deck.

Furniture

- Sofas, chairs, tables, lamps, accessories, and artwork. A large comfortable lounge chair with an ottoman has been requested.
- Provide a reading lamp.
- Provide numerous floor and sofa pillows. The floor pillows are for Marissa to recline with for a nap or while watching children's videos.

Design

- Carpet floor covering is required.
- Provide selection of wall and ceiling finishes.
- Provide custom detailing for bases and trims.
- Provide fireplace elements, including front facing, mantle, hearth, and screen, as well as materials and finishes.
- Incorporate a lighting solution for the Living Room that carries through to the Dining Room and provides general illumination, and features artwork. Provide dimmers for the general illumination.
- Window treatments are required.

Equipment

- Fireplace with gas starter
- Television, VCR, DVD and stereo equipment.
- Cordless telephone

DINING ROOM

Planning

- Adjacency: Living Room and Kitchen
- The dining room is used frequently for large family gatherings and entertaining.
- This space is to be an open plan and have direct access to the Living Room and Kitchen.
- They need a table that will seat ten to twelve people. Larger parties will be served buffet style.
- Storage is required for china, serving tableware, silver, and linens.

Furniture

- Dining table
- Dining chairs for twelve people
- Buffet and server
- Table accessories, artwork, and other furnishings you find appropriate for the space

Design

- Preference is for an expandable glass top table.
- Carpet floor covering is required.
- Provide selection of wall and ceiling finishes.
- Incorporate a lighting solution that will provide general illumination, dining task lighting, and feature artwork. Provide dimmers for the general illumination.
- Window treatments are required.
- Provide custom detailing for bases and trims.

KITCHEN

Planning

- Adjacency: Dining Room
- Provide counter area with four or five stools for casual dining, buffet serving, or bar setup. One section is to be handicap accessible.
- Develop a functional and convenient arrangement around a "work triangle".[1]
- Provide a double stainless steel sink and dishwasher area with counter space on each side.
- Provide a refrigerator/freezer and adjacent counter area.
- Provide a cooking area with counter space on each side. Cooktop must include a griddle. Provide proper ventilation.
- Provide wall and base cabinets for storage of kitchen linens, food, cookbooks, utensils, bowls, storage ware, small appliances, dishes, flatware, glassware, cookware and bake ware. Provide a pot rack, built-in wine and stemware racks. Some shelving should be open for display of special serving ware.
- The following cabinet features have been requested: pullout drawer units at all base cabinets, "Lazy Susan" racks in all corner cabinets, extra pullout work surface in solid surface material, drawer dividers for cutlery and flatware, towel bars under the sink cabinet, cutting board door racks, door-mounted spice racks, and recycle trash receptacles.

Design

- Hard surface flooring is required; no ceramic tile.
- Provide selection of wall and ceiling finishes.
- Provide custom detailing for bases and trims.
- Design an area for a small, potted herb garden.

---

1. Designing Interiors, p. 224.

- Study all design transition elements and finishes as they relate to adjacent areas. Use logical solutions to material changes.
- Window treatment is required.

Equipment

- Refrigerator, dishwasher, stainless steel cooktop, hood, stainless steel oven, and microwave
- Double stainless steel sink and disposal
- Telephone
- Fire extinguisher

BATHROOM

Planning

- Adjacency: Marissa's Bedroom and Guest Bedroom
- Both guests and Marissa will use the bath.
- Provide storage space for items related to grooming, bath items, and linens.

Furniture

- Mirrors, bath accessories, door hooks, towel bars, artwork, and area carpets

Design

- Hard surface floor covering is required.
- Provide selection of wall and ceiling finishes.
- Provide custom detailing for bases and trims.
- Provide specialty fixtures and casework.

Equipment

- Lavatory, vanity, water closet, and bathtub/shower combination

MARISSA'S BEDROOM

Planning

- Adjacency: Bathroom
- The primary use is as a sleeping area.
- Marissa will also require a play area and storage for clothing, toys, and large collection of stuffed animals.
- Her favorite character is Barney and she especially loves solving puzzle games. She already has an intense appetite for fashion.

Furniture

- Full–size bed with removable child safety rails and related furniture
- Child's table and chairs
- Toy storage

Design

- They do not want a juvenile theme design in the room.
- Carpet floor covering is required.
- Provide selection of wall and ceiling finishes.
- Provide custom detailing for bases and trims.
- Window treatments are required.

Equipment

- Future telephone and computer

GUEST BEDROOM

Planning

- Adjacency: Bathroom
- This bedroom is for overnight guests and will serve as a study area for Marissa when she gets older.
- Provide clothing storage closet area.

Furniture

- Full- or double-size bed with related furnishings

Design

- Carpet floor covering is required.
- Provide selection of wall and ceiling finishes.
- Provide custom detailing for bases and trims.
- Provide custom bedding and window treatment.

Equipment

- Telephone and future computer

MASTER BEDROOM

Planning

- Locate in a separate zone of the first level. Adjacency: Master Bathroom and Deck
- The room will be an important meeting place for Charles and Mitzi to come together for conversation at the end of the day. As a peaceful sanctuary for them, it is primarily a sleeping area but will also function as a private retreat for relaxing, watching television, or reading.
- Plan a private sitting area on the Deck.

Furniture

- King–size bed and related furniture
- Personal storage
- Full-length mirror
- Comfortable lounging chair and related furniture
- Lamps, accessories, artwork, and other appropriate furnishings

Design

- A large window frames the outdoor landscape so the design is to be played down to emphasize this view. The color combination of ivory and black has been requested, with ivory as the predominant color.
- Custom bedding is required.
- Carpet floor covering is required.
- Provide selection of wall and ceiling finishes.
- Provide custom detailing for bases and trims.
- Window treatments with blackout capability are required.

Equipment

- Ceiling- or wall-mounted television
- Cordless telephone

MASTER BATH

Planning

- Create a sense of flow from the sleeping area into the master bath.
- Provide storage space for items related to grooming, personal effects, bath items, and linens.

- Provide two separate grooming areas.
- Provide an area with a make-up vanity and chair in Mitzi's grooming area.
- Provide a walk–in custom marble shower.
- Provide an enclosed water closet area.

Furniture

- Mirrors, bath accessories, door hooks, towel bars, artwork, and area carpets
- Any furnishings you find appropriate

Design

- Provide finished marble floor, shower floor, and whirlpool bath enclosure. Glass block wall shower enclosure is required. Design a seat in the shower.
- Provide selection of wall and ceiling finishes.
- Provide specialty fixtures and casework.
- Provide custom detailing for cabinets, bases, and trims.
- Window treatments are required.

Equipment

- Two lavatories, water closet, whirlpool bath, and double showerhead with additional wall water jets
- Telephone

MASTER CLOSET

Planning

- Separate closets are preferred but are not required.
- This dressing area requires storage for hanging and folded clothing, shoes, and accessories. A total of 40 linear feet is satisfactory.
- Create separate hanging sections (of equal lengths) for suits, dresses, jackets, and shirts.
- Provide bins for shoes and sweaters.
- Provide eight drawers with dividers for undergarments, socks, and accessories.
- Provide a portable hamper.

Design

- Relate to the bedroom scheme.

LAUNDRY

Planning

- Plan for convenience and function, including necessary appliances and fixtures.
- Provide a counter for folding clothes.
- Provide storage for laundry and cleaning supplies.

Design

- Provide selection of wall and ceiling finishes.
- Provide shelving and/or cabinet storage, bases, and trims.
- Hard surface flooring is required.

Equipment

- Washing machine, clothes dryer, and laundry sink
- Telephone

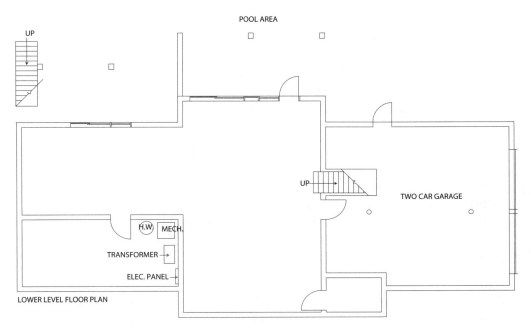

POOL AREA

UP

UP

TWO CAR GARAGE

H.W   MECH.

TRANSFORMER →

ELEC. PANEL →

LOWER LEVEL FLOOR PLAN

Figure 22–4   Lower Level Floor Plan

DECK

Planning
- Plan seating arrangements for conversation and dining and a lounging group.
- Plan a small area for grill cooking. The main grill center is on the lower patio.

Furniture
- Settees, chairs, ottomans, chaise lounges, tables, and planting pots

Design
- All materials must be suitable for outdoor use.
- Provide a retractable awning to shade the deck and block the sunlight from the interior.

Equipment
- Exterior stereo speakers

Lower Level Spatial Requirements

FAMILY ROOM

Planning
- Adjacency: Outdoor Patio and Pool
- Another multipurpose room, both family members and guests will spend a significant amount of time in the family room. Of prime importance is a home entertainment area that incorporates a media wall center and can seat eight to ten people.
- Charles and Mitzi hope to entertain often around the pool and want a small wet bar with storage for liquor, soft drinks, and glassware incorporated. This facility will be a future kitchenette.
- Plan an area for a folding game table and chairs for four people for playing cards or games.
- Provide an area for playing pool and electronic darts.
- Provide an area for aerobic exercise equipment.
- Provide secondary seating arrangements of bar chairs and small, high tables adjacent to the pool table area.

Furniture

- Sofas, chairs, ottomans, tables, lamps, accessories, and artwork
- Étagère to feature family photos, Charles's trophies from motor cross races, and fashion books
- Charles has four large photographic prints of Indy cars, a wheel, and a steering wheel that he wants to show at the game area and pool table.

Design

- Having so many functions, it is important to create a sense of harmony within the spaces.
- Carpet floor covering is required.
- Provide selection of wall and ceiling finishes.
- Provide custom detailing for bases and trims.
- General illumination, accent lighting, and pool table pendant light are required. Provide dimmers for the general illumination.
- Window treatments are required.

Equipment

- Complete home entertainment system, including television and VCR and DVD equipment. Integrate the equipment into the overall design scheme.
- Bar: sink, refrigerator, and ice machine
- Exercise equipment that includes a stepping machine, ski machine, and treadmill
- Telephone

OFFICE

Planning

- This room will be used exclusively as a home office for Mitzi and Charles.
- Provide two desk areas, storage for mail, work-related items, file storage, and book storage. The desks may be built in or a free standing furniture configuration.

Furniture

- Desk-height work surfaces
- File storage
- Book shelves
- Two desk chairs
- Lamps, waste receptacles, accessories, and artwork. The artwork is to incorporate photos of Mitzi and her awards and projects.

Design

- Carpet floor covering is required.
- Provide selection of wall and ceiling finishes.
- Provide custom detailing for bases and trims.

Equipment

- Computer, and combined printer, fax, and copier
- Telephone

GUEST BATHROOM

Planning

- Both guests and family will use the bath.
- Provide storage space for items related to grooming, bath items, and linens
- Bath is to be completely accessible, including a roll-in shower.

Furniture

- Mirrors, bath accessories, door hooks, towel bars, and artwork

Design

- Hard surface floor covering is required.
- Provide selection of wall and ceiling finishes.
- Provide custom detailing for bases and trims.
- Provide specialty fixtures and casework.

Equipment

- Lavatory, vanity, water closet, and shower

WINE CELLAR/STORAGE

Planning

- Plan an area for wine storage. Four linear feet will be sufficient.
- This room is also used for cold storage, so two linear feet of hanging and shelving are required.
- Allow room for folding table and chair storage.
- Other storage requirements are for bicycles, snow and water skis, and seasonal decorations.
- Include an area for a small safe.

Design

- Hard surface floor covering is required.
- Provide selection of wall and ceiling finishes.
- Provide custom detailing for wine racks, storage units, bases, and trims.

Equipment

- Safe
- Wine cooling system

POOL PATIO

Planning

- Plan a seating arrangement for conversation, a dining area, and lounging groups for sunning.
- Plan an outdoor bar area and large grill center.
- Provide a portable cabana/gazebo to block equipment on the right side of the pool.

Furniture

- Settees, chairs, ottomans, chaise lounges, tables, umbrellas, toy storage, and planting pots

Design

- Materials must be suitable for outdoors.

SUBMITTAL REQUIREMENTS

Phase I: Programming   Part A Due: _____   Part B Due: _____

Part C Due: _____

    A. Analyze the program outlined above and develop a programming matrix keyed for adjacency locations. A sample blank matrix is provided in Chapter 31.

    B. With the information gathered from the matrix, provide a minimum of three bubble diagrams for the spatial relationships of the areas required.

    C. Provide a one-page typed concept statement of your proposed planning direction and initial design concept.

    Presentation formats for Parts A through C: White bond paper, 8½″ × 11″, with a border. Insert in a binder or folder. The matrix and bubble diagrams are to be completed on the CADD program.

    Evaluation Document: Programming Evaluation Form

Phase II: Conceptual Design   Date Due: _____

    A. Develop preliminary plans based on the functional and aesthetic requirements necessary, including use of space, furniture layouts, and fixtures.

    Evaluation Document: Conceptual Design Evaluation

Phase III: Design Development   Date Due: _____

    A. Based on the approved preliminary plans, develop and present for review and approval, final interior design recommendations to establish and describe the character of the interior space of the project.

    B. Prepare documents and illustrations in the form of plans, elevations, four to six color and material sample boards, photographs of furnishings and fixtures, and two color renderings to show:

        a. Wall, furniture, and fixture arrangement
        b. Floor, wall, window, and ceiling treatments
        c. Furnishings, fixtures, and millwork including custom trims
        d. Schematic lighting plan and lighting fixtures
        e. Colors, materials, and finishes
        f. Suggestions for art, accessories, and plants

    Evaluation Document: Design Development Evaluation Form

Phase IV: Final Design Presentation   Date Due: _____

    A. The final presentation will be an oral presentation to a panel jury. Project must be submitted on 20″ × 30″ illustration boards. Your name, the date, the project name and number, the instructor's name, and the course title must be printed on the back of each board in the lower left-hand corner. Submit as many boards as required, including renderings, elevations, fixture and furniture selections, material selections, and plans.

    B. The plans should be presented in 1/4″ scale. The elevations are to be presented in 1/2″ scale.

    Evaluation Document: Final Design Presentation Evaluation Form

Phase V: Construction Documentation   Date Due: _____

    A. Prepare working drawings and specifications for non-load-bearing interior construction, materials, finishes, furnishings, fixtures, and equipment for the clients' approval. The drawings and specifications must include:

       1. A furniture plan.

       2. Plan indicating any special design elements and attached interior fixtures keyed to details, sections, and elevations as required.

       3. Elevations or details of walls showing design elements, fixtures, and color and material indications keyed to details and finish specification schedules.

       4. A floor covering plan and details keyed to specification schedules.

       5. Details showing sections through specially designed elements, custom trim profiles or fixtures must be prepared and identified as to wall locations.

       6. A lighting criteria plan must be provided and keyed to a lighting fixture schedule.

       7. An electrical power criteria plan must be provided and keyed to a legend. Plans or elevations are to include stereo speakers and smoke detectors.

       8. A reflected ceiling plan indicating new ceiling conditions, materials, and details with specifications must be prepared. Plans or elevations are to include stereo speakers and smoke detectors.

       9. All materials and finishes must be specified on color and material schedules.

    B. The plans should be presented in 1/4″ scale. The elevations are to be presented in 1/2″ scale.

Evaluation Document: Construction Documentation Evaluation Form

Project Time Management Schedule: Schedule of Activities, Chapter 31.

The AutoCAD LT 2002 drawing file name is hahn.dwg and can be found on the CD.

REFERENCES

Trade Magazines

    Interior Design Buyers Guide: http://www.interiordesign.net

Special Resources

    See ADA references in Chapter 31.

Product Manufacturers

    Abet Laminati: http://www.abetlaminati.com
    B & B Italia: http://www.bebitalia.it
    E15 Design: http://www.e15.com
    Euro Tile: http://www.euro-tile.com
    Formglas: http://www.formglas.com
    Formica Corporation: http://www.formica.com
    Jacuzzi, Inc.: http://www.jacuzzi.com
    KraftMaid: http://www.kraftmaid.com
    Leucos, Inc.: http://www.leucos.com
    Marimekko: http://www.marimekko.com
    M2L: http://www.m2lcollection.com

Rainier Richlite Co.: http://www.richlite.com
Robert Allen: http://www.robertallendesign.com
Scavolini Kitchens: http://www.scavolini.com
Vinotemp International: http://www.vinotemp.com
Waterworks: http://www.waterworks.com
Wilsonart International: http://www.wilsonart.com

NOTES

CHAPTER 23

Da-Shi Villa

OUTCOME SUMMARY

The project focuses on developing conceptual skills for the design of residential facilities, working in metric scale, and project management documentation through the use of control sheets. You will be required to conceptualize a design image that is heavy with architectural details and elaborate materials. The project features three phases of the design process: identifying and analyzing the client's needs and goals, developing conceptual skills through schematic or initial design concepts, and detailing and refining ideas from the schematic design phase.

client profile

Your clients, Ms. Lu and Mr. Yao, a lovely couple living in Taiwan, are exceptionally affluent and prominent individuals. Teddy, who is 10 years old, is their only child. In meeting Ms. Lu, you will without a doubt, find her thoughtful, poised, and especially chic. Given that she has an around-the-world-life, Ms. Lu is a fusion of the various influences she has encountered. This reality enters into her design attitudes. Her husband, Mr. Yao, has a very enthusiastic nature and owns a major corporation located in Kaohsiung, a key industrial hub of Taiwan.

They are building a country home north of Taipei called Da-shi Villa situated on a lake with a 27-hole golf course nearby. They would like you to design and orchestrate the entire image of the villa. This includes plan layouts of the furnishings, furniture, color and material recommendations, elevations and architectural details, renderings, and, finally, documentation for Jack Shen, the architect in Taiwan. Mr. Shen will complete the construction documents and bid the project. They expect a time-honored Western or European look that is stunning and comfortable and does not incorporate traditional Asian décor. Well-appointed furnishings and luxurious materials are a few key features that are important to them.

Jack has requested that you meet in New York City to review the preliminary concepts and look at furnishings at the design centers. This requires that you do a bit of legwork beforehand researching the showrooms that would be suitable to visit during your short time in the city. Plan one day to meet and review your schematic preliminary design presentation and the second day to tour the showrooms. Prepare an agenda for the meeting. The client and consultants understand, but speak very little English; so Mr. Shen has arranged for a Chinese interpreter.

As the site survey photographs show, the home is well under construction and the walls are firmly in place. You may well even say, "Cast in stone!" Concrete is a characteristic building material in this area and used widely for home construction as well. With the walls in place, your most important planning tasks include the furniture layouts. This needs to be done quickly and expediently in order to verify that all power requirements are properly in place.

Guests are treated to a multilevel experience when they enter the villa, and the high, slanted ceilings present a real lighting and wall design challenge. Fortunately, there are many windows and skylights flooding natural daylight into the space. The first level is an open plan with the exception of the kitchen and laundry. A more formal dining experience is important to their mode of entertaining and they have asked that the transition into the dining room be softened with architectural elements. Tripanel wood doors are already at the construction site and are to be used throughout the villa. The kitchen has been completed by others and is not part of your design contract.

The location is in an earthquake zone, and seismic precautions and guidelines are to be applied to life safety issues.

project details

General and Architectural Information

Location: Taichung, Taiwan

Gross square meters: Three Stories: 465. First Level: 164. Second Level: 123. Lower Level: 178. (Approximately 5,000 square feet gross building area)

The architect/contractor will administer construction documents and the bidding of the interior furniture, fixture, and finish work.

New construction Architectural plans are available. Start construction in February.

Consultants required: Architect, electrical engineer, and mechanical engineer

Codes and Governing Regulations

Like California, Taiwan is an earthquake region and special life safety issues apply to construction. The UBC contains information from the seismological and structural engineering communities to reduce the risk of damage from earthquakes and windstorms.

Occupancy: Residential. Single-Family Dwelling

Design the project to comply with the most current edition of the International Residential Code (IRC). The minimum habitable room size is 120 square feet. The minimum ceiling height for habitable rooms is 7′ 0″. The minimum handrail height is 34″. Smoke alarms must be installed in every bedroom, any hallway leading to a bedroom, and on each level of a multistory building. Safety bars on windows must release easily in an emergency. Windows and hallways should be clear of obstructions. Use great judgment when placing large pieces of furniture in front of windows that are an escape route. The Seismic Zones have been replaced by IRC Seismic Design Categories. There are detailed provisions for wall bracing, including construction methods, connection requirements, anchorage, and seismic requirements.

Facilities Required

First Level: Entry, Stairs, Living Room, Sitting Room, Dining Room, Guest Bathroom, and Patio. (Kitchen, Laundry and Storage Rooms are not in the design contract.)

Upper Level: Balcony and Hall, Master Bedroom, Master Bath, Master Closet, Teddy's Bedroom, Guest Bedroom, and Bathroom

Lower Level: Entertainment Room, Game Room, Bathroom, and Maid's Bedroom. (Storage Room is not in the design contract.)

First Level Spatial Requirements

ENTRY

Planning

- The entry is the main doorway for greeting guests and the only transition space between each level.

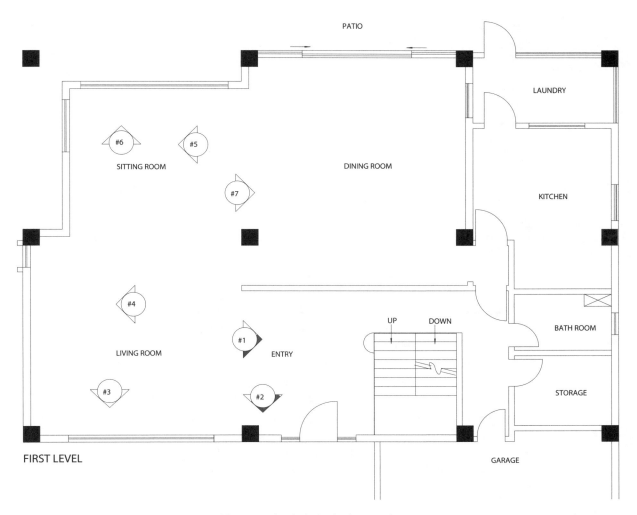

PATIO

#6

#5

SITTING ROOM

#7

DINING ROOM

LAUNDRY

KITCHEN

#4

LIVING ROOM

UP   DOWN

#1

ENTRY

BATH ROOM

#3

#2

STORAGE

FIRST LEVEL

GARAGE

Figure 23–1   First level plan with photograph notations

Figure 23–2   Photo 1: Entry

Figure 23–3   Photo 2: Entry

- Develop a furniture plan that directs circulation and gives a good feeling upon entering.
- Provide special storage for shoes that are removed when entering the home.

Furniture

- Console-type table
- Any other furnishings that you find appropriate
- Plants, accessories, and artwork
- Area carpet

Design

- Make a leading design statement in the space that will resonate throughout the first level and transitional spaces.
- Provide a marble floor throughout the first level that has elaborate custom details. This material is to continue into the Living Room, Sitting Room, Dining Room, hallways, and Bathroom.
- Create a concept for stair carpeting and area carpets.
- Provide selection of wall and ceiling finishes.
- Provide custom detailing for bases, trims, and column treatments.
- The architecture of the ceiling is very dramatic in this space and special lighting and design considerations are required.
- Design the finish detailing for the stairs, including tread and riser materials, handrails, balusters, newel posts, and balcony treatments. The location and strength of these architectural elements make them a strong focal point of the Entry and Living Room. Because of their visibility, these fundamentals should also reaffirm the design personality of the interior on the whole.
- Select wood finish paint or stain color for the Entry doors and passage doors.
- Incorporate chandeliers and wall sconces as part of the lighting solution for the Entry and Stairs.

LIVING ROOM

Planning

- This room is used primarily to entertain guests and should be an atmosphere favorable to conversation.

Figure 23–4 Photo 3: Living Room

Figure 23–5 Photo 4: Living Room

- The fireplace is the architectural focal point of the room.
- Create an intimate seating arrangement around the fireplace to accommodate approximately eight to ten people.

Furniture
- Sofas, chairs, tables, lamps, accessories, area carpets, and artwork

Design
- Provide window treatments.
- Provide selection of wall and ceiling finishes.
- Provide custom detailing for bases, trims, and column treatments.
- The architecture of the ceiling is very dramatic in this space and special lighting and design considerations are required.
- Provide fireplace elements, including front facing, mantle, hearth, and screen, as well as materials and finishes.

SITTING ROOM

Planning
- This area is a secondary "social space" where family will gather together or be alone to watch television, read, or relax.
- Ms. Lu also will use it for recreational reading or to simply rest her head, for solace and quiet.
- Provide a comfortable and intimate furniture grouping for the family.
- Create adequate definition between this area and the Dining Room.

Furniture
- A suitable enclosure for the television and equipment
- Sofas, chairs, tables, lamps, accessories, area carpets, and artwork

Design
- Ms. Lu expressed a desire to possibly use furnishings designed by some of the prominent fashion designers in this space, specifically the Fendi or Versaci Collections.
- Window treatments are required but cannot completely block the natural light or view through the back windows.

Equipment
- Television and VCR/DVD equipment.
- Telephone

Figure 23–6 Photo 5: Sitting Room

Figure 23–7 Photo 6: Sitting Room

DINING ROOM

Planning

- As noted above it is important to create definition between this area and the sitting room with architectural detailing, floor details, ceiling details, and lighting.
- The Dining Room is used for family dining and is also used extensively for entertaining guests, both formally and informally.
- They have requested a round table that will seat twelve people. A rotating center section in the table-top must also be incorporated for traditional Chinese serving.
- Storage is required for china, tableware, silver, glassware, and linens.
- The sliding doors open to the patio, allowing dinner guests to easily transition to the outdoors.

Figure 23–8   Photo 7: Dining Room

Furniture

- Custom dining table
- Dining chairs for twelve people
- Buffet and server
- Lamps, accessories, area carpets, artwork, and any other furnishings that you find appropriate

Design

- They would like the room to have highly detailed architectural elements. Include special column treatments, moldings, ceiling coves, and a prominent central feature above the chandelier.
- This room is to incorporate special detailing in the marble floor areas as well, to help define the room perimeter.
- The have asked for Regency-style furniture for this room.
- There is a service window to the laundry that requires careful visual consideration.
- Provide custom detailing for bases, trims, and column treatments.
- Window treatments are required.

GUEST BATHROOOM

Planning

- This is the only bath on this level. Both guests and family members will use it.
- Provide storage for linens.

Furniture

- Bath accessories, mirrors, area carpets, artwork, and any other furnishings that you find appropriate

Design

- Provide marble floor and walls.
- Incorporate furnishings and/or design elements that will relate to the mood of the first level.
- Provide specialty fixtures and casework.
- Provide custom detailing for trim and casework fixtures.

Equipment

- Lavatory, vanity, water closet, and bathtub/shower combination

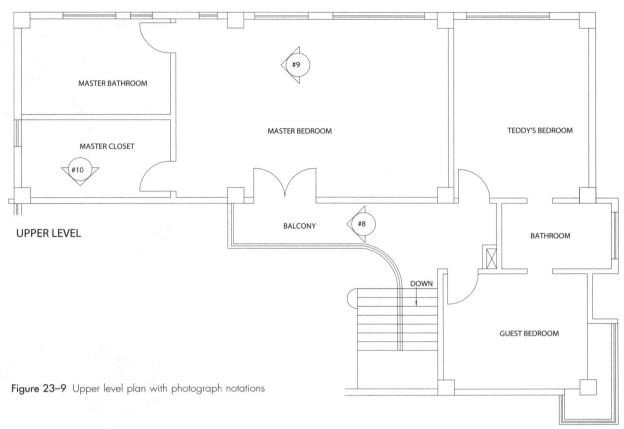

UPPER LEVEL

Figure 23–9 Upper level plan with photograph notations

PATIO

Planning

- Plan a seating arrangement for six people for conversation, a dining area, and lounging group.

Furniture

- Settees, chairs, ottomans, chaise lounges, tables, umbrellas, and planting pots

Design

- Fabric-upholstered seat cushions suitable for outdoors are required.

Upper Level Spatial Requirements

BALCONY AND HALL

Planning

- In this transition space between the social and private areas of the home, circulation is the primary function.
- Sight lines are important because of the visibility from the first level.

Furniture

- Any furnishings that you find appropriate
- Wall sconces
- Area carpets

Design

- Carry through the design concept from the lower level, including balcony-railing treatment.
- Provide marble flooring.

Figure 23–10
Photo 8: Balcony, looking down into Living Room

MASTER BEDROOM

Planning

- This is the bedroom for Ms. Lu and Mr. Yao and their private retreat for sleeping, relaxing, watching television, reading, or listening to music.
- It is primarily a sleeping area.
- A second function requires a sitting area for reading or watching television.
- A third requires a desk area for correspondence and telephone.

Furniture

- King-size bed and related furniture
- Personal storage
- Desk with chair
- Comfortable lounging chairs or sofa grouping and related furniture
- Entertainment unit for television and stereo
- Lamps, accessories, mirrors, area carpets, artwork, and any furnishings you find appropriate

Figure 23–11
Photo 9: Master Bedroom

Design

- The architecture of the ceiling is very dramatic in this space and special lighting and design considerations are required.
- The client prefers an upholstered headboard and requested custom bedding.
- Marble floor and/or combination of marble and carpeting is required.
- Provide window treatments.

Equipment

- Television and stereo
- Telephone

MASTER BATH

Planning

- Provide storage space for items related to grooming and bath linens.

Furniture

- Bath accessories, mirrors, area carpets, artwork, and plants
- Any furnishings that you find appropriate for the space

Figure 23–12
Photo 10: Master Bedroom Loft Space

Design

- Provide marble floor and walls and tub and shower.
- Specialty fixtures and casework
- Shutters on windows are required.

Equipment

- Two lavatories, water closet, whirlpool bath, and shower

MASTER CLOSET

Planning

- There is a loft above this walk-in-type closet. Develop ideas for its use and access to it.
- This dressing area requires storage for hanging and folded clothing, shoes, and accessories.

Furniture

- A dressing table with chair
- Mirror

Design

- Relate to the bedroom scheme.

Equipment

- Note that there is an ironing center built into the wall.

TEDDY'S BEDROOM

Planning

- The primary use is a sleeping area.
- Teddy will require a study area.
- The closet area is also a transitional space into the Bathroom.

Furniture

- Full bed and related furniture
- Mirror, accessories, and artwork

Design

- They do not want a childlike theme design in the room.
- Fabric covering or wood veneer is required on closet doors. Provide a visually pleasing solution for the closet/passage into the Bathroom.
- Carpet floor covering is required.

Equipment

- Television
- Computer
- Telephone

GUEST BEDROOM

Planning

- This bedroom is used exclusively for overnight guests.
- Plan for their sleeping comfort and other overnight needs of luggage and clothing storage.

Furniture

- Full- or double-size bed with related furnishings
- Mirror, accessories, artwork, and plants

Design

- Carpet floor covering is required.
- Custom bedding and window treatment are required.

Equipment

- Telephone

BATHROOM

Planning

- This bath will be used by Teddy and overnight guests and has access from both bedrooms.
- Provide storage for linens, Teddy's grooming items, and guest items.

Furniture

- Bath accessories, mirrors, area carpets, and artwork

Design
- Provide marble floors and walls.
- Specialty fixtures and casework

Equipment
- Lavatory, water closet, and combination bath and shower

Lower Level Spatial Requirements

ENTERTAINMENT ROOM

Planning
- Both family members and guests spend a significant amount of time in the entertainment room. Of prime importance is a home theater area that incorporates a media wall center. Plan to seat eight to ten people.
- Plan for a fully functioning bar area to seat two. Provide 4 to 6 linear feet of storage for fine wines.
- Plan an area for a table and chairs to seat four people.
- Provide a secondary seating arrangement of lounge chairs adjacent to the bar area.
- The wall elements are to incorporate extensive use of full-height storage units, including shelving and cabinets. All walls are to have a wood, raised panel application.
- A dry garden is required adjacent to the stairs that is approximately the width of the stairs.

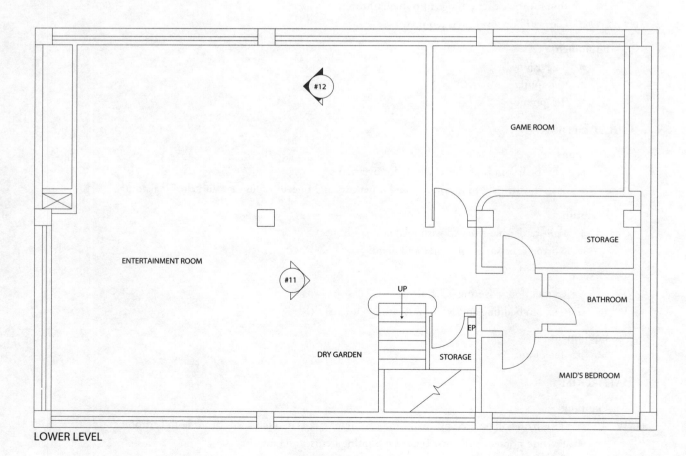

Figure 23–13 Lower level plan with photograph notations

Furniture

- Sofas, chairs, tables, lamps, bar stools, accessories, and artwork
- Area carpets

Design

- Create interest with the built-ins to make the room more intimate.
- Provide a combination of hard surface and carpet flooring.
- Provide window treatments.
- Provide selection of wall and ceiling finishes. They have requested a drywall and molding crafted coffered ceiling treatment.
- Provide custom detailing wall, cabinets, bar, bases, and trims.

Equipment

- Complete home entertainment system including television, VCR and DVD equipment, and stereo sound system. This equipment is to be housed in custom wall units and hidden behind cabinet doors to integrate it into the overall design.
- Bar: sink, wine cooler, refrigerator, dishwasher, and ice machine
- Telephone

GAME ROOM

Planning

- Provide custom-built wall storage for games on one wall and a table seating arrangement for four to play mah-jong and other board or card games.
- Provide an area for Teddy to play video games.

Furniture

- Table and chairs for four people

Design

- Provide wood finished walls.
- Provide adequate lighting at the game table.

Equipment

- Television and video game system
- Telephone

Figure 23–14 Photo 11: Entertainment Room

Figure 23–15 Photo 12: Entertainment Room

BATHROOM

Planning

- This is the only bath on this level and is used by guests, family members, and the maid.
- Provide storage for linens.

Furniture

- Bath accessories, mirrors, area carpets, and artwork

Design

- Hard surface floor covering and walls are required.
- Provide specialty fixtures and casework.

Equipment

- Lavatory, vanity, water closet, and bathtub/shower combination

MAID'S BEDROOM

Planning

- Plan an efficient, compact space for resting and sleeping.

Furniture

- Daybed, storage for clothing, desk, chair, side table, and lamp

Design

- Provide window treatments, floor covering, and paint finishes.

SUBMITTAL REQUIREMENTS

Phase II: Conceptual Design   Date Due: _____

A. Develop preliminary plans based on the functional and aesthetic requirements necessary, including furniture layouts and fixtures.

B. Prepare an agenda for client meeting.

Evaluation Document: Conceptual Design Evaluation Form

Phase III: Design Development   Date Due: _____

A. Based on the approved preliminary plans, develop and present for review and approval final interior design recommendations to establish and describe the character of the interior space of the project.

B. Prepare documents and illustrations in the form of plans, elevations, four to six color and material sample boards, photographs of furnishings and fixtures, and two color renderings to show:

a. Wall, furniture, and fixture arrangements
b. Floor, wall, window, and ceiling treatments including enlarged details of custom patterns
c. Furnishings, fixtures, and millwork
d. Schematic lighting plan and lighting fixtures
e. Colors, materials, and finishes
f. Suggestions for art, accessories, and plants

Evaluation Document: Design Development Evaluation Form

Phase IV: Final Design Presentation   Date Due: _____

A. The final presentation will be an oral presentation to a panel jury. Project must be submitted on 20″ × 30″ illustration boards. Your name, the date, the project name and number, the instructor's name, and the course title must be printed on the back of each board in the lower left-hand corner. Submit as many boards as required, including renderings, elevations, fixture and furniture selections, material selections, and plans.

B. The plans should be presented in 1:50 metric scale. Elevations should be presented in 1:25 metric scale.

Evaluation Document: Final Design Presentation Evaluation Form

Phase V: Construction Documentation   Date Due: _____

A. Prepare a project record sheet for each room.

B. Prepare working drawings and specifications for non-load-bearing interior construction, materials, finishes, furnishings, fixtures, and equipment with enough information to submit to the architect for final drawings. The drawings and specifications must include:

1. Plans indicating any special design elements and attached interior fixtures keyed to details, sections, and elevations as required.
2. Elevations or details of walls showing design elements, fixtures, and color and material indications keyed to details and control sheets.
3. Furniture plans keyed to control sheets.
4. Floor covering plans and details keyed to control sheets.
5. Details showing sections through specially designed elements or fixtures must be prepared and identified as to wall locations.
6. Lighting criteria plans must be provided and keyed to a lighting fixture schedule.
7. Electrical power criteria plans must be provided and keyed to a legend. Plans or elevations are to include stereo speakers and smoke detectors.
8. Reflected ceiling plans indicating new ceiling conditions, materials keyed to control sheets, and details must be prepared. Plans or elevations are to include stereo speakers and smoke detectors.
9. All materials and finishes must be specified on the control sheets.

C. The plans should be presented in 1:50 metric scale. Elevations should be presented in 1:25 metric scale.

Evaluation Document: Construction Documentation Evaluation Form

Project Time Management Schedule: Schedule of Activities, Chapter 31.

Project Record Sheet Form: Project Record Sheet, Chapter 31.

The AutoCAD LT 2002 drawing file name is da-shi.dwg and can be found on the CD.

REFERENCES

Trade Magazines

Interior Design Buyers Guide: http://www.interiordesign.net

Resources.com: http://www.resources.com

Special Resources

Quake Safety: http://www.qsafety.com

Product Manufacturers

Accessories International: http://www.accintl.com

Alistair Mackintosh Ltd.: http://www. alistairmackintosh.co.uk

Artimide : http://www.artemide.com

Artistic Frame: http://www.artisticframe.com

Avery Boardman: http://www.averyboardman.com

Cozzolino Decorative Wall Panels: http://www.cozzolino.com

Donghia: http://www.donghia.com

Flos: http://www.flos.net

Frederick Cooper Lamps: http://www.frederickcooper.com

Hartco Wood Flooring: http://www.hartcoflooring.com

Jeffco: http://www.jeffcofurniture.com

John Saladino: http://www.saladinofurniture.com

Kallista Kitchen & Bath: http://www.kallista.com

Karges Furniture Company: http://www.karges.com

LaBarge: http://www.labargeinc.com

Mirak Furniture: http://www.mirakfurniture.com

Modern Stone Age, Inc.: http://www.modernstoneage.com

Pallas Walls: http://www.pallastextiles.com

Sherle Wagner: http://www.sherlewagner.com

Stark Carpet Corporation: http://www.starkcarpet.com

Steve Harsey Textiles: http://www.stevenharsey.com

Valli & Valli: http://www.vallievalli.com

NOTES

CHAPTER 24

Cruz Residence

OUTCOME SUMMARY

The project features three phases of the design process: identifying and analyzing the client's needs and goals, developing conceptual skills through schematic or initial design concepts, and detailing and refining ideas from the schematic design phase. The project focuses on increasing knowledge of design practice for single-family homes, and design for infant and toddler spaces.

client profile

Al and Biana Cruz, a sophisticated Southport couple, have hired you to fuse formal and casual interior design elements while keeping an eye on tradition and comfort for their new home. Delicate balances of formal and casual spaces abound throughout. Some of the home's features include high sloping ceilings at the entry, great room, and master bedroom. Sliding doors and lanky windows stretch across the back of the home on the first level importing light and opening the space to the outdoors. Doors open onto the deck from both the kitchen and rear corridor, allowing free flow movement to the outside.

Biana, a woman of taste and style, is clever at bringing together beautiful things in a comfortable setting. She wants everyone to feel good when they walk into a room in her home. She has been saving magazine clippings of spaces and items she loves, most of which show interiors with soft, light colors and eclectic furnishings. Ceiling fans are to be installed in the kitchen, great room, and bedrooms. Comfortable, enjoyable, connected, pleasurable, welcoming, and practical are the catchwords for their environment. They have great family gatherings during the holiday season with more than thirty adults and their children participating in the festivities. Al and Biana have small, intimate cocktail parties with their friends. They plan to entertain them in the great room and the wine cellar.

Al started his professional life as an entrepreneur. He founded both a culinary school and technical institute, which he developed, and profitably sold a few years later. Shortly thereafter he started another successful business venture. Biana's area of expertise is marketing and human resources. Together they are a dynamite team. Both are passionate about winter sports and they travel to the mountains to snow ski whenever they can.

They are recently married, with their first child, Colin, on the way. They plan to have more children in the near future and can count on having twins since they both come from large families with several sets of twins, including Biana. To ensure the safety of the children, specify electrical outlet covers, child safety locks on low cabinets, medicine and household chemical supply storage, and provide safety bars for the windows. In addition provide child safety window covering hardware, nonskid mats under area carpets, and unlock protection levers on interior doors. The children's rooms must function to span various ages therefore; you should plan for appropriate age group conversions.

project details

General and Architectural Information

Gross square footage: Three Stories. 5,200. First Level: 2,500. Upper Level: 1,700. Lower Level: 1,000.

Project budget: $140,000 furniture budget. The architect/contractor will administer construction documents and the bidding of the interior furniture, fixture, and finish work.

New construction. Architectural plans are available.

Consultants required: Architect, electrical engineer, and mechanical engineer

Codes and Governing Regulations

Occupancy: Residential, Individual One-Family Dwelling

Design the project to comply with the most current edition of the International Residential Code (IRC) and the Americans with Disabilities Act. The minimum habitable room size is 120 square feet. The minimum ceiling height for habitable rooms is 7′ 0″. The minimum handrail height is 34″. Smoke alarms must be installed in every bedroom, any hallway leading to a bedroom, and on each level of a multistory building. Safety bars on windows must release easily in an emergency. Windows and hallways should be clear of obstructions. Use great judgment when placing large pieces of furniture in front of windows that are an escape route.

Facilities Required

First Level: Entry, Living Room, Dining Room, Entry Powder Room, Den/Study, Kitchen, Laundry, and Great Room

Second Level: Master Bedroom, Mater Bath, Master Closet, Nursery, Toddler's Bedroom, Children's Bedroom, and Bathroom

Lower Level: Wine Cellar

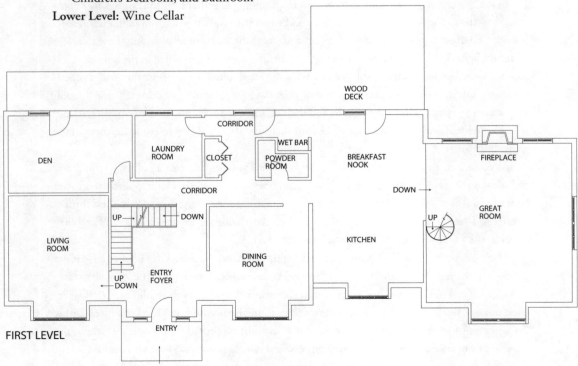

Figure 24–1 First Level Floor Plan

First Level Spatial Requirements

ENTRY

Planning

- The Entry is the main entrance for guests only and functions as a passage space between the Living Room, Dining Room, and upper and lower levels. There are several circulation corridors to the other areas of the residence as the plan shows.
- Develop a plan that directs circulation and creates a sense of interconnectedness.

Furniture

- A center hall table or wall console with artwork or mirror is requested by the owners.
- Any furnishings that you find appropriate for the space
- Plants, area carpets, artwork, and accessories, including a pair of very large urns or vases flanking the entryway and fish pots

Design

- Create a design that graciously welcomes guests with elegance and warmth.
- Provide a hard surface floor material throughout the first level entry and corridors. Carefully consider its transition into the adjacent rooms.
- Create a concept for stair carpeting and area carpets.
- Select wood finish paint or stain color for the tread and riser, handrails, balusters, newel posts, entry doors, and passage doors.
- Provide custom detailing for bases and trims.
- Provide gates for toddler safety at all stairs.
- Incorporate a lighting solution for the Entry including general illumination, a pendant light, and accent lighting for artwork.
- Study all design transition elements and finishes as they relate to adjacent rooms and the hallways. Use logical solutions to material changes.

Figure 24-2 Entry window wall

Figure 24-3 Entry stairs

Figure 24-4 Entry walls

LIVING ROOM

Planning

- This room is used solely to entertain guests and should be an atmosphere favorable to conversation.
- Provide a comfortable and intimate seating arrangement to accommodate approximately six to eight people.

Furniture

- Sofas, chairs, tables, lamps, accessories, and artwork

Design

- Incorporate current elements of great style and sophistication.
- Provide carpet floor covering.
- Provide selection of wall and ceiling finishes and door hardware.
- Provide custom detailing for bases and trims.
- Provide soft general illumination on dimmers and accent lighting for artwork.
- Window treatments are required.

DINING ROOM

Planning

- The dining room is used infrequently for family dining, business dinners, and entertaining guests formally.
- The owners require a table that will seat ten people. Larger parties will be served buffet style.
- Storage is required for china, tableware, silver, glassware, and linens.
- Plan a serving area.

Furniture

- Dining table with double leaf and table pads
- Fully upholstered host and hostess chairs
- Dining chairs for eight people
- Buffet, china cabinet, and server
- Chandelier, accessories, artwork, and any other furnishings that you find appropriate for the space

Design

- Provide carpet floor covering.
- Provide selection of wall and ceiling finishes. Biana wants to try a deep red paint with a suede look finish.
- Provide custom detailing for bases and trims.
- Provide soft illumination on dimmers and accent lighting for artwork. Provide a pendant fixture above the dining table and wall-mounted fixtures on the buffet wall.
- Window treatments are required.

ENTRY POWDER ROOM

Planning

- Both guests and family members will use the bath.
- Provide storage for linens.

Furniture

- Bath accessories, hook on door, task lighting, towel bars, mirrors, and artwork
- Step stool for children

Design

- Relate to Entry design elements.
- Provide solid surface countertops and backsplash with seamless sink.
- Hard surface floor is required.
- Provide selection of wall and ceiling finishes and door hardware.
- Provide custom detailing for bases and trims.
- Lighting must incorporate wall and ceiling fixtures.
- Provide specialty fixtures and casework.

Equipment

- Lavatory, vanity, and water closet

DEN/STUDY

Planning

- This is used primarily as a home office for tasks, such as correspondence, reading, and computer use.
- Plan storage units and work surfaces to function as an office. Provide a desk area, computer area, storage for work-related items, file storage, book storage, and television viewing.

Furniture

- Desk that accommodates standard paperwork duties with pencil and small box drawers
- Minimum twenty linear feet of bookshelves
- Large distressed leather lounging chair, ottoman, and related furnishings
- Lamps, area carpets, accessories, and artwork

Design

- The owners have requested an eclectic casual and comfortable look that includes traditional elements.
- Provide wood flooring.
- Provide selection of wall and ceiling finishes and door hardware.
- Provide custom detailing for bases and trims.
- Window treatment is required.

Equipment

- Computer, printer, fax, and copier; separate lines for telephone, computer, and fax machine
- Telephone
- Small television

KITCHEN

Planning

- Develop a functional and convenient arrangement around a "work triangle."[1]
- Provide a seamless solid surface double sink. Dishwasher area (left-handed) should have counter space on each side. Food disposal must be located in the sink.
- Provide a Sub Zero refrigerator and adjacent counter area.
- Cooking island area with counter space on each side is required. Cooktop must include modular interchangeable elements for the grill. Provide proper ventilation. Island is to be multi-height for food preparation and dining.

---

1. *Designing Interiors*, p. 224.

- Provide wall and base cabinets for storage of linens, food, cookbooks, utensils, bowls, storage ware, small appliances, serving ware, dishes, flatware, cutlery, glassware, cookware, and bake ware.
- The following cabinet features have been requested: two built-in wood cutlery dividers, tilt-out storage at sink cabinet, two multistorage pantry cabinets with in-door storage, three special storage cabinets for larger serving ware and bake ware, and four deep and tall drawers for cookware.
- Dining countertop at the island should seat three. This area will be used for casual dining.
- Provide a casual dining area with a table and chairs to seat six people.
- To run the daily functions of the household, provide a telephone/work desk area with seating and filing and storage capabilities.
- Provide a wet bar in the adjacent corridor.

Design

- Provide wood surface flooring.
- Provide selection of wall and ceiling finishes and door hardware.
- Solid surface countertops and backsplash with seamless sink in the kitchen are required.
- Provide custom detailing for bases and trims.
- Cherry wood cabinets and overhead glass rack and shelves at the wet bar are required.
- Window treatments are required.
- Provide both general illumination and plenty of task lighting. Provide a pendant fixture with a fan over the kitchen table.

Furniture

- Three bar-height stools
- Table and dining chairs
- Desk chair
- Area carpets, accessories, and artwork

Equipment

- Refrigerator, dishwasher, cooktop, double oven, microwave, bread warmer, and trash compactor
- Double sink and disposal
- Brass sink and fixtures in wet bar
- Portable telephone
- Fire extinguisher

LAUNDRY

Planning

- Plan for convenience and the functions of sorting, washing, drying, and folding clothes, including necessary appliances, counters, and fixtures.
- Provide a deep laundry sink area with counter space and storage on each side.
- Provide an ironing center work area.
- Plan storage for house cleaning equipment and supplies.
- Plan storage for small household tools, batteries, flashlights, lightbulbs, etc.

Furniture

- Counter stool
- Window treatment
- Area carpets

Design
- Hard surface flooring is required.
- Provide selection of wall and ceiling finishes and door hardware.
- Provide shelving and wall and base cabinets, bases, and trims.

Equipment
- Washing machine, clothes dryer, and deep laundry sink
- Wall mounted telephone

GREAT ROOM

Planning
- Family members will spend a significant amount of time in the Great Room playing with children, watching television, and being with friends and family for casual entertainment. Seat eight to ten people in a primary seating group.
- Plan an area with a table and chairs to seat four to six people.
- Provide a secondary seating arrangement of lounge chairs with ottomans.
- Provide a play area and toy storage under the spiral staircase. Design a playful wall fixture that includes a low table for a train set or dollhouse, shelving, and storage cabinets accessible to toddlers and children. When the children are older, their primary play area will be in the lower level Game Room and this area should be planned for a future piano.

Furniture
- Sofas, chairs, tables, lamps, accessories, plants, and artwork
- Cabinet for VCR and DVD equipment
- Portable storage for approximately forty DVDs and VCR tapes
- Large floor pillows
- Fireplace accessories

Design
- Carpet floor covering is required.
- Window treatments are required.
- Provide selection of wall and ceiling finishes.
- Provide custom detailing for bases and trims.
- The architecture of the ceiling is very impressive in this space and special lighting and design considerations are required. Provide a pendant light at the table area.
- Provide fireplace elements, including front facing, mantle, hearth, and screen, as well as materials and finishes.

Equipment
- Large-screen (thirty six inch) television
- VCR and DVD equipment
- Portable telephone

Second Level Spatial Requirements

MASTER BEDROOM

Planning
- It is primarily a sleeping area but will also function as a private retreat for relaxing, watching television, or reading.
- The secondary function requires a sitting area for Biana to read, make telephone calls, or use for late night or early morning baby feedings.

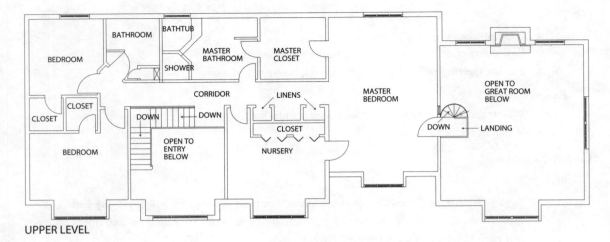

Figure 24–5 Upper Level Floor Plan

Furniture

- King-size bed with a deep mattress and related furniture
- Personal storage
- Bench for sitting while putting on shoes
- Comfortable lounging chairs and related furniture
- Lamps, accessories, artwork, and any other appropriate furnishings

Design

- Carpet floor covering is required.
- Provide custom bedding, including pillow shams and throw pillows.
- Provide selection of wall and ceiling finishes and door hardware.
- Provide custom detailing for bases and trims.
- Window treatments with blackout capability are required.

Equipment

- Television
- Portable telephone

Figure 24–6 Master Bedroom

MASTER BATH

Planning

- Provide storage space for items related to grooming, bath items, and linens.

Furniture

- Mirrors, bath accessories, and door hooks, and any furnishings that you find appropriate

Design

- Provide hard surface floor and walls and tub and shower. Shower enclosure must be made of clear glass. Design a seat in the shower.
- Provide selection of wall and ceiling finishes and door hardware.
- Provide custom detailing for cabinets, bases, and trims.

Figure 24–7   Master Bath Shower

Figure 24–8   Master Bath Whirlpool Bath

Equipment
- Specialty fixtures and casework
- Two lavatories, water closet, whirlpool bath, and showerhead with additional wall water jets

MASTER CLOSET

Planning
- This dressing area requires storage for hanging and folded clothing, shoes, and accessories.
- Incorporate a built-in ironing center.

Design
- Relate to the bedroom scheme.

NURSERY

Planning
- Plan for twins.
- This is the sleeping area for the children while they are infants. There is access from both the Master Bedroom and hallway.
- Provide a changing table area, sleeping area, feeding area with rocker, and small play area.
- Provide bins, drawers, shelving, and hang bars in closet area.

Furniture
- Cribs, rocker, changing table, mirror and related furnishings
- Diaper disposal and hamper
- Lamps, accessories, ceiling mobiles, and artwork

Design
- Carpet floor covering is required.
- Provide selection of wall and ceiling finishes and door hardware.
- Incorporate a window seat with storage for books and infant toys.
- Window treatments are required.

Equipment
- Baby monitor

TODDLER'S BEDROOM

Planning
- This is the sleeping area that will remain a toddler's bedroom until the children are older.
- Plan for a twin bed sleeping area.
- Provide a play area with small lounge chairs for reading and toy storage.

- Storage should be easily accessible for the children.
- Provide a safe way for children to escape in case of fire.

Furniture

- Juvenile-scale furniture
- Two twin beds with safety rails and related furnishings
- Clothing storage and toy storage chests
- Dresser with mirror
- Small-scale lounge chair
- Lamps, hamper, accessories, and artwork

Design

- Carpet floor covering is required.
- Provide selection of wall and ceiling finishes and door hardware.
- Concept must be suitable for both sexes and not be a juvenile theme.
- Window treatments are required.

Equipment

- Baby monitor

CHILDREN'S BEDROOM

Planning

- The primary use is as a sleeping area for children 4 years and older.
- A study area is also required.
- Storage should be easily accessible for the children.
- Provide a safe way for children to escape in case of fire.

Furniture

- Beds for two children and related furniture
- Lamps, accessories, and artwork

Design

- Provide selection of wall and ceiling finishes and door hardware.
- They do not want a juvenile theme design in the room.
- Carpet flooring is required.
- Window treatments are required.

Equipment

- Computer
- Telephone

BATHROOM

Planning

- The bath is used for the grooming and bathing of infants.
- The children will use this bath as well.
- This bath is used as a temporary location for potty training for children.
- Provide storage for linens, grooming items, and infant grooming needs.
- Incorporate a pullout hamper in the base cabinet.

Furniture

- Child's stool at each lavatory
- Accessories, mirrors, towel bars and hooks accessible to children, diaper disposal, artwork, and area carpets

Design

- Provide hard-surface floors (nonskid) and walls.
- Provide selection of wall and ceiling finishes and door hardware.
- Provide specialty fixtures and casework.
- Window treatments are required.

Equipment

- Lavatory for bathing infants, water closet, and combination bath and shower with nonskid floors

Lower Level Spatial Requirements

WINE CELLAR

Planning

- This room is used for entertaining, wine storage, and cigar storage.
- Plan for both red and white wine storage and wine serving. Provide a built-in cabinet with 4 to 6 linear feet of storage for fine wines, and a wine cooler.
- Provide a seating area for six with a table. Biana prefers an upholstered bench or sofa on one side of the table and stools, chairs, or a church pew on the other side.
- Provide storage for cigars in a humidor cabinet.

Furniture

- Table with seating for six
- Bar stools
- Two lounge chairs and related furniture
- Lamps, area carpets, accessories, and artwork

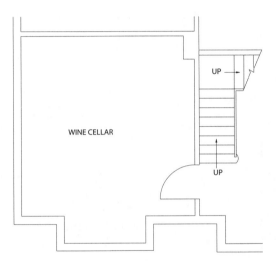

Figure 24–9   Lower Level. Wine Cellar Plan

Design

- Provide selection of wall and ceiling finishes and door hardware.
- Provide soft general illumination, task lighting, and accent lighting.
- Wall units providing wine storage and counter surface area are required. Include additional storage in base cabinets.
- Provide hard surface flooring.

Equipment

- Wine cooling system
- Humidor

SUBMITTAL REQUIREMENTS

Phase II: Conceptual Design   Date Due: _____

A. Develop preliminary plans based on the functional and aesthetic requirements necessary, including furniture layouts and fixtures.

Evaluation Document: Conceptual Design Evaluation Form

Phase III: Design Development   Date Due: _____

A. Based on the approved preliminary plans, develop and present for review and approval final interior design recommendations to establish and describe the character of the interior space of the project.

B. Prepare documents and illustrations in the form of plans, elevations, a minimum of six color and material sample boards, photographs of furnishings and fixtures, and two color renderings to show:

a. Wall, furniture, and fixture arrangement
b. Floor, wall, window, and ceiling treatments
c. Furnishings, fixtures, and millwork
d. Schematic lighting plan and lighting fixtures
e. Colors, materials, and finishes
f. Suggestions for art, accessories, and plants

Evaluation Document: Design Development Evaluation Form

Phase IV: Final Design Presentation   Date Due: _____

A. The final presentation will be an oral presentation to a panel jury. Project must be submitted on 20″ × 30″ illustration boards. Your name, the date, the project name and number, the instructor's name, and the course title must be printed on the back of each board in the lower left-hand corner. Submit as many boards as required, including renderings (quantity to be determined by instructor), elevations, fixture and furniture selections, material selections, and plans.

B. The plans should be presented in 1/4″ scale. The elevations are to be presented in 1/2″ scale.

Evaluation Document: Final Design Presentation Evaluation Form

Project Time Management Schedule: Schedule of Activities, Chapter 31.

The AutoCAD LT 2002 drawing file name is cruz.dwg and can be found on the CD.

REFERENCES

Trade Magazines

Interior Design Buyers Guide: http://www.interiordesign.net

Product Manufacturers

Avonite, Inc.: http://www.avonite.com
Bruce Hardwood Floors: http://www.bruce.com
Century Furniture: http://www.centuryfurniture.com
Corian Solid Surfaces: http://www.corian.com
Cox Furniture Company: http://www.coxmfg.com
CTH Sherrill Furniture: http://205.160.234.84/
David Edward: http://www.davidedward.com
Decorative Crafts: http://www.decorativecrafts.com
Dennis Miller Associates: http://www.dennismiller.com
DesignTex: http://www.dtex.com
Donghia Furniture and Textiles Ltd.: http://www.donghia.com
Duravit Living Bathrooms: http://www.duravit.com
Hickory White: http://www.hickorywhite.com
Kohler: http://www.kohler.com
Lexington Furniture: http://www.lexington.com
Maitland-Smith: http://www.maitland-smith.com
Nevamar Decorative Surfaces: http://www.nevamar.com
Ohio Table Pad Company: http://www.otpc.com
Sarreid: http://www.sarreid.com
Sherrill Furniture: http://www.sherrillfurniture.com
Vinotemp International: http://www.vinotemp.com
Vitraform: http://www.vitraform.com

CHAPTER 25

Van Paten Residence

OUTCOME SUMMARY

The project features three phases of the design process: identifying and analyzing the client's needs and goals, developing conceptual skills through schematic or initial design concepts, and detailing and refining ideas from the schematic design phase. The project focuses on increasing knowledge of design practice for a large single-family home, planning for security and safety, and preparing a preliminary budget. Brainstorming exercise is included to develop an awareness of safe rooms and shelter room interior requirements and responsibilities.

client profile

Annabel and Raul Van Paten are a charismatic young couple with a delightful countenance that includes strong roots of spirituality and great generosity. To the Van Patens their home is a place to be with family and be inspired by their surroundings. Their fondness for music ranges from Bach to Bon Jovi and the children can dance and mimic the entire repertoire.

The Van Patens have many goals for the interiors of their roomy palazzo-like home in Hilton Head, South Carolina. They want you to make the spacious home warm, cozy and comfortable, as well as creating rooms that are ideal for upscale entertaining. They have frequent family gatherings of twenty people or more for birthdays, holidays, and special occasions. Small intimate get-togethers with six to eight friends occur on occasion. They host many out-of-town friends as overnight guests. The Van Patens prefer a traditional look with elaborate architectural details and warm, rich colors mixed with neutrals. Create a design with a feeling of opulence and warmth, because they love wood in medium and dark hues and generous carved shapes and details. The heart of family living for this couple and their children will be the open Great Room area, Kitchen, and Family Room. Layouts must be compatible with easy living as well as accommodating the needs of children and parents equally.

Much of Annabel's time involves taking care of their children, Sarah, 4, and John, 2, so her elementary education teaching degree comes in very handy. It is evident that it has been put to good use because the children are bright and advanced in skill development. They are musically inclined and love to sing and dance. Sarah is quite attached to her stuffed characters, Tweedy and Bunny, and three dolls, all called Baby. Toy storage is needed in each room where the family spends time together. Annabel is tough when she must be, nurturing when she needs to be, and always decisive and charming. She has a wide-ranging porcelain figurine and crystal collection just as glamorous looking as she is. She wants a perfect place to help the groups stand out.

Figure 25–1 Exterior Front

Figure 25–2 Exterior Rear

Raul is "Mr. Hospitality" because he has a wonderful natural ability for making you feel welcome and comfortable in his home. He is a grand host who is also a brilliant businessman. For more than 15 years Raul has owned and directed a family business that was started by his father over 50 years ago. He is involved in many national organizations, and has spent a great deal of time traveling and networking over many rounds of golf before he started his family.

The exterior of the home is terra cotta colored brick with tan window trim and fascia with gray sandstone detailing. Segmental raised brick accentuates the corners. The interior construction is wood framing and studs, with gypsum board wall and ceiling finishes. The residence has many window walls that ensure the interior makes the most of outer views. You can take pleasure in the landscape from nearly any perch in the home. The entrance hall has soaring windows above the doors opening to a full sky view, and to the swimming pool in the rear. The abundance of natural light welcomes you into the home with warmth and grace. The many high vaulted ceiling spaces enhance the sense of drama. Walls of French doors punctuate the entrance hall, Great Room, and Master Bedroom, that enable movement to the outside from many of the spaces and almost blur the distinction between inside and outside spaces.

Total home security devices will be installed, including access control devices and intrusion detection devices, which are tied into the local police authorities. Complete house intercom system with monitors and whole house stereo speaker system with controls in each room as well as exterior speakers at the pool will be installed. Special consideration is to be given in the design of both a safe room and shelter room for the home in the event of intrusion and emergency. Provide fire extinguishers and smoke and carbon monoxide detectors throughout the home.

project details

General and Architectural Information

Gross square footage: 10,800. Two-story single-family residence with basement. First Level: 5,500. Upper Level: 2,500. Lower Level: 2,800

Project budget: The owner requires the interior designer to develop a preliminary furniture budget. The owner will administer construction documents and the bidding of the interior furniture, fixture, and finish work.

Opening or move-in date: December 10th

New construction. Architectural plans are available.

Consultants required: Architect, electrical engineer, mechanical engineer, and home security specialist

Figure 25–3 First Level Floor Plan

Codes and Governing Regulations

Occupancy: Residential, Individual One-Family Dwelling

Design the project to comply with the most current edition of the International Residential Code (IRC). The minimum habitable room size is 120 square feet. The minimum ceiling height for habitable rooms is 7′ 0″. The minimum handrail height is 34″. Smoke alarms must be installed in every bedroom, any hallway leading to a bedroom, and on each level of a multistory building. Safety bars on windows must release easily in an emergency. Windows and hallways should be clear of obstructions. Use great judgment when placing large pieces of furniture in front of windows that are an escape route.

Facilities Required

First Level: Foyer, Sitting Room, Dining Room, Guest Bathroom, Family Room, Office, Kitchen, Pool Bath, Great Room, Master Bedroom, Master Bath, Master Closet, Nursery, Nursery Bath, and Patio

Upper Level: Upper Foyer, Guest Bedroom, Guest Bathroom, Bedroom 1, Bedroom 2, Children's Bathroom, Linen Room, Laundry, and Playroom and Exercise Room

Lower Level: Club Room, Bathroom, Laundry/Sewing Center, Baking Kitchen, Storage Room, and Shelter Room

First Level Spatial Requirements

FOYER

Planning

- The Foyer is the primary access for guests and is the main circulation corridor on the first level as well as the transition space between the lower level social spaces and upper level private zones.
- Develop a furniture plan that supports the needs of family and guests in addition to directing circulation.
- The Foyer closet must have storage for coats, hats, gloves, scarves, umbrellas, diaper bags, camcorders and camera equipment, and central vacuum attachments.

Figure 25–4   Entrance Hall looking toward the front entry

Furniture

- Any furnishings that you find appropriate
- Chandeliers, plants, accessories, and artwork
- Create a concept for stair carpeting and area carpets

Design

- The sizable entrance hall, with a ceiling high enough to fly a kite, has oversized Palladian windows and French doors at the rear, which fill the vaulted space with light.
- Provide a hard surface floor material. The transition from the Foyer into the Family Room, Kitchen, Sitting Room, and Dining Room is one step down and must be clearly defined for safety.
- Provide selection of wall and ceiling finishes.
- Provide custom detailing for bases and trims.
- Design the finish detailing for the stairs, including tread and riser materials, handrails, balusters, newel posts, and balcony treatments. The location and power of these architectural elements make them a strong focal point of the Foyer.
- Select front and rear entry door systems.
- Select wood finish paint or stain color for the entry doors and passage doors.
- Incorporate a lighting solution for the Foyer that will provide general illumination, create a soft glow in the evening, and add accent lighting to feature artwork.
- Study all design transition elements and finishes as they relate to adjacent rooms. Use logical solutions to material changes.

Equipment

- House intercom panel

SITTING ROOM

Planning

- The family will use this room rarely. Most of its use will occur during large gatherings and during the holiday season. Oftentimes the Christmas tree will be located here in front of the window. Provide a comfortable and intimate seating arrangement to accommodate approximately six to eight people.

- Create an atmosphere favorable to conversation.

Furniture

- Sofas, chairs, tables, curio cabinet, lamps, accessories, and frames for the large portraits of the children

Design

- Provide carpet floor covering.
- Provide selection of wall and ceiling finishes.
- Provide custom detailing for bases and trims.
- Window treatments are required.

DINING ROOM

Planning

- The dining room will be used for entertaining guests formally and for large family birthday celebrations and occasional holiday dinners.
- They need a table that will seat twelve people. Larger parties will be served buffet style.
- Storage is required for china, other tableware, silver, glassware, and linens.

Furniture

- Dining table
- Dining chairs for twelve people
- Buffet, china cabinet, and server
- Lamps, chandeliers, accessories, area carpet, artwork, and any other appropriate furnishings

Design

- Provide hard surface floor covering.
- Provide selection of wall and ceiling finishes.
- Provide custom detailing for bases and trims.
- Window treatments are required.

GUEST BATHROOM

Planning

- Both guests and family members will use the bathroom.

Furniture

- Mirrors, lamps, chandeliers, bath accessories, door hooks, area carpet, artwork, and any other appropriate furnishings

Design

- Provide hard surface floor covering.
- Provide selection of wall and ceiling finishes.
- Provide custom detailing for bases and trims.
- Specialty fixtures and casework

Equipment

- Lavatory, vanity, and water closet

FAMILY ROOM

Planning

- Both family members and guests will spend a significant amount of time in the Family Room. Seating for six to eight people is required. Family members will typically use it when an extra TV is needed. It will also be used for cozy entertaining of small gatherings.
- Provide storage for toys within the furnishings.
- Incorporate a wall cabinet to house the television set and the main stereo equipment for the home. Include storage for all types of audio and visual tapes.

Figure 25–5   Entrance Hall looking into the Family Room

- Plan for a fully functioning wet bar area. Seating is not required at the bar. Provide 2 linear feet of storage for fine wines, storage for liquor, soft drinks, bar ware, and six types of glassware.

Furniture

- Sofas, chairs, tables, curio cabinet, lamps, magazine storage, accessories, and artwork
- Fireplace accessories

Design

- Provide carpet flooring. Hard surface flooring is required at the bar area.
- Provide selection of wall and ceiling finishes.
- Provide custom detailing for bases and trims.
- Window covering is required.
- The fireplace is to take center stage in the room. Design fireplace elements, such as front facing, mantle, hearth, and screen, as well as materials and finishes.
- Provide a custom cabinet for the bar area, and television, stereo, and related equipment.

Equipment

- Portable telephone
- Television and stereo system in wall cabinet
- VCR and DVD equipment
- House intercom panel
- Bar sink, wine cooler, and bar refrigerator

Figure 25–6   Family Room looking toward the Entrance Hall

Figure 25–7   Family Room window wall

OFFICE

Planning

- This room will be used very frequently as a home office. Raul will study paperwork and read. The children will play some computer toddler games, and Raul has some favorite computer games that require steering wheel and other game hardware. Research and shopping via the Internet is also a frequent pastime.
- Plan storage for books, files, stationery, and computer equipment. Provide a work area and a reading area.

Furniture

- Executive desk with file storage, executive chair, two large leather lounge chairs, lamps, accessories, and artwork

Design

- Raul has requested a warm oak finish and wishes to incorporate a traditional wood crafted coffered ceiling.
- Provide hard surface wood floor and raised wall panels.
- Provide selection of wall and ceiling finishes.
- Provide custom detailing for bases and trims, and double French entry doors with glass.
- Window treatment is required.

Equipment

- Wireless Media Center
- Computer station, printer, fax, copier
- Telephone. Separate the phone lines for telecommunication equipment and high-speed Internet cable access
- House intercom panel

KITCHEN

Planning

- The Kitchen best tells the story of how this family lives and fosters togetherness. Raul likes to grab a quick cup of coffee and watch the morning news before he leaves for the office. The children will watch morning television shows while they eat their breakfast at the counter.
- The Van Patens would like a large island with room for prep work and seats for socializing.
- Develop a functional and convenient arrangement around a "work triangle"[1] that includes the cooking island.
- Provide a double stainless steel sink and dishwasher area with counter space on each side. Provide deep sink basins to accommodate large pots. Multispray, pullout faucet is required.
- Provide a Sub Zero refrigerator and a Sub Zero freezer and adjacent counter area.
- Cooking island area with counter space on each side is required. Provide proper ventilation for a gas cooktop.
- The following cabinet features have been requested: under counter pullout base cabinets, pots and pans drawers, sliding shelves in base cabinets, four cutlery dividers, wide drawer dividers, and "lazy Susans" in the corner cabinets.
- Wall and base cabinets for storage of linens (two drawers), food, cookbooks, utensils (four drawers), bowls (one cabinet), storage ware (one cabinet), small appliances, serving ware (two cabinets), dishes (two cabinets), flatware (one drawer), glassware (three cabinets), cookware (five drawers), and bake ware (two cabinets).

---

1. Kilmer, p. 224.

- Provide a countertop at the island for eating with two bar stools. This area will be used for casual dining and buffet serving.
- Incorporate a message board by the telephone with room for a calendar and address book and telephone book storage close by.
- Provide a casual, easy-going dining area with a table and chairs for six people. This dining space is to be located next to the window to take advantage of the natural light. The area is used daily for family lunch and dinner.
- Provide shelving in the separate pantry area for food and appliance storage because they have just about every small appliance available for food preparation.
- The children will spend time playing while Annabel cooks so toy storage is also required.

Furniture

- Table and chairs for six people
- Decorative buffets, chandelier, accessories, and artwork
- Two counter-height swivel stools
- Step stool for the pantry

Design

- Provide selection of wall and ceiling finishes.
- Provide hard surface flooring.
- Provide granite countertops and backsplash.
- Provide custom detailing for bases and trims.
- Lighting solution must include general illumination, chandeliers, task lighting, and accent lighting.
- Window treatment is required.

Equipment

- Refrigerator, freezer, dishwasher, cooktop, ventilation, two ovens with convection cooking, and microwave
- Double sink and disposal
- Portable telephone
- Small countertop television with cable hookup
- Fire extinguisher
- House intercom panel
- Security system panel

POOL BATH

Planning

- Plan a private changing area for guests with custom built-in storage for clothing, pool towels, pool life jackets, and toys.
- Plan a private bathroom area, including a water closet and lavatory.
- Include a concealed laundry center. Plan for convenience and function, including necessary appliances, counters, ironing center work area, and fixtures.

Furniture

- Vanity chair and bench
- Wall towel hooks for pool towels and door hooks
- Accessories, area carpets, and artwork

Design

- Hard surface flooring is required.

- Provide selection of wall and ceiling finishes.
- Provide a custom built-in vanilty-like table with mirror and personal storage.
- Provide custom detailing for bases and trims.

Equipment
- Water closet and lavatory
- Washing machine, clothes dryer, and laundry sink

GREAT ROOM

Planning
- Both family members and guests will spend a significant amount of time in the Great Room, and a high level of comfort should be provided, including huge leather sectionals and generous tables.
- Plan a seating arrangement for ten to twelve people.
- Plan an area for two table groups that each seat four people. It will be used for dining and playing cards or games.
- Provide toy storage.
- One wall is to incorporate use of custom storage units, including shelving and cabinets for a large-screen television, VCR and DVD equipment, books, photo albums, large videotape collection including family videos, and display for accessories and family photographs.

Furniture
- Sofas, chairs, tables, lamps, accessories, and artwork. The large family portrait will be hung over the fireplace.
- Two to three chests and/or large drawers for toy storage.
- Fireplace accessories

Design
- The room is very large with church-like proportions, and the furniture and design elements must help to define the space and create intimacy.
- Carpet floor covering is required.
- Provide fireplace elements, such as front facing, mantle, hearth, and screen, as well as materials and finishes. Window covering is required.
- The lighting should be soft overhead type that creates an atmosphere for peaceful relaxation in the evening.

Equipment
- Large-screen, 73″, digital television
- VCR and DVD equipment
- House intercom panel
- Portable telephone

MASTER BEDROOM

Planning
- This is a luxurious space that includes a window wall and a wall of French doors leading to the patio and pool area.
- It is primarily a sleeping area but will also function as a private retreat for relaxing, meditation, watching television, or reading in bed. Raul is the early riser.
- Annabel has requested a desk and chair area for personal correspondence.

Furniture

- King-size bed and related furniture
- Wardrobe for television and personal storage
- Writing desk with storage for stationery and writing supplies and a chair
- Lamps, accessories, artwork, and any other appropriate furnishings

Design

- Provide custom bedding.
- Carpet floor covering is required.
- Provide selection of wall and ceiling finishes.
- Provide custom detailing for bases and trims.
- Window treatments with blackout capability are required.

Equipment

- Television
- Telephone
- House intercom panel
- Security system panel

Figure 25–8   Master Bedroom window wall

MASTER BATH

Planning

- Plan a large walk-in shower with a clear glass shower enclosure. Design a seat for the shower.
- Provide storage space (15 drawers) for items related to grooming, bath items, and eight linear feet of shelving for linens .
- Plan a vanity seating area for Annabel.
- This will serve as a temporary location for potty training for children.

Furniture

- Vanity chair
- Bath accessories, door hooks, artwork, area carpets, and any appropriate furnishings

Design

- Hard surface floor and wallcovering and custom tub and shower enclosure are required.
- Provide selection of wall and ceiling finishes.
- Lighting must include general illumination, adequate task lighting for grooming, and a heat lamp near the shower entry.
- Provide custom detailing for cabinets, bases, and trims.

Equipment

- Specialty fixtures and casework
- Two lavatories, water closet, bidet, whirlpool jet bath, and two showerheads with an additional four wall water jets for the shower area
- Telephone
- House intercom panel
- Small television

MASTER CLOSET

Planning

- This dressing area requires storage for hanging and folded clothing, shoes, and accessories.
- Provide two dressing station areas that incorporate mirrors, three drawers, and lighting.
- Create separate hanging sections along two walls for suits, dresses, jackets, and shirts.
- Provide bins for shoes and sweaters.
- Provide storage for folded clothing along one wall, with twelve drawers with dividers for undergarments, socks, and accessories, and a three-way mirror.
- Provide lowheight shelving and drawers in the center of the space.
- Incorporate a built-in ironing center.

Furniture

- Bench for sitting while putting on shoes
- Three-way mirror
- Hamper

Design

- Relate to the bedroom scheme.
- Provide window covering.

NURSERY

Planning

- This is the sleeping area for the children while they are infants and toddlers.
- Provide a changing table area, sleeping area, feeding area with rocker, and small play area.
- Provide sufficient space for free play and many large stuffed animals.
- Sarah has an angel collection that she wants next to her bed.

Furniture

- Crib and toddler bed, clothing storage, rocker, changing table, minimal toy and book storage, diaper storage, mirror, and related furnishings
- Existing rocking horse
- Lamps, diaper waste receptacle, accessories, and artwork

Design

- Annabel is excited about using very bright wall colors in the space.
- Carpet floor covering is required.
- Provide selection of wall and ceiling finishes.
- Bookcase and cabinet for storage of books and infant toys and a storage chest are required.
- Window treatments are required.

Equipment

- House intercom panel
- Baby monitor

NURSERY BATH

Planning

- The bath is used for the grooming and bathing of infants.
- The children will use this bath as well.
- Provide storage for bath toys, linens, grooming items, and infant grooming needs.
- Incorporate a pullout hamper in the base cabinet.

Furniture

- Two child's stools for reaching sink
- Bath accessories, artwork, and area carpets

Design

- Hard surface floors and walls are required.

Equipment

- Specialty fixtures and casework
- Lavatory with baby washing features, water closet, and combination bath and shower. Incorporate child safety features and provide a showerhead that can be adjusted for the children.

PATIO

Planning

- Provide an area enclosure, circulation, and play space.
- Plan a seating arrangement for conversation, a dining area, and a lounging group. They will entertain up to twenty people.

Furniture

- Settees, chairs, ottomans, chaise lounges, tables, umbrellas, and planting pots
- Umbrella for sand box and sand toy storage
- Doll house furniture (child's scale) including a table, two chairs, rocking chair, pillows for loft bed and fabric selection for window covering.
- Storage for pool floats, shuffleboard equipment, and lawn bowling games in an out of sight area
- Two water hose containers

Design

- Provide a pool enclosure.
- Provide a shuffleboard court.
- Fabric-upholstered seat cushions suitable for the outdoors are required.
- Provide awnings.
- Provide selection of patio and walkway pavers or floor surface and pattern.

Upper Level Spatial Requirements

The Van Patens, keen on safety for their family, want you to determine which room on this level to use as a safe place to wait out an intrusion or emergency as part of their home safety plan. The room, sometimes referred to as a "panic room," must have safe and secure doors and windows, and include a cell phone, fire extinguisher, bottled water, flashlight, batteries, pepper spray, first aid kit, 911 light, fire escape ladder, portable radio and/or a television, and some type of diversion for the children at minimum. In addition, the room is to be airtight and also used for any disasters that may compromise the integrity of the air in the home system or exterior elements.

UPPER FOYER

Planning

- The upper foyer is the transition space into the private bedroom areas and has visual exposure from the lower Foyer and Family Room.
- Develop a furniture plan that directs circulation as well as continuing the design concept from the main Foyer.

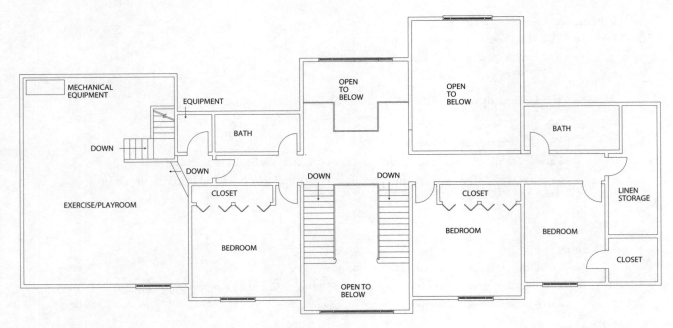

Figure 25–9   Upper Level Floor Plan

Furniture

- Any furnishings that you find appropriate
- Plants, accessories, and artwork

Design

- Provide carpet floor covering.
- Provide selection of wall and ceiling finishes.
- Provide custom detailing for bases and trims.
- Incorporate a lighting solution for the entry that will provide general illumination and accent lighting for artwork.
- Study all design transition elements and finishes as they relate to adjacent rooms and the hallway to the private zones of the home. Use logical solutions to material changes.

GUEST SUITE
GUEST BEDROOM

Planning

- This bedroom is used exclusively for overnight guests.
- Plan for their sleeping comfort and other overnight needs of luggage and clothing storage.
- Plan a seating area or writing desk area.

Furniture

- King-size bed with related furnishings
- Any furnishings that you find appropriate
- Plants, accessories, and artwork

Design

- Carpet floor covering is required.
- Provide custom detailing for bases and trims.
- Provide custom bedding and window treatment.

Equipment

- Television
- Telephone
- House intercom panel

GUEST BATHROOM

Planning

- Overnight guests will use this bathroom.
- Provide storage for linens and grooming items.

Furniture

- Bath accessories, door hooks, artwork, and area carpets

Design

- Hard surface floor covering is required.
- Provide selection of wall and ceiling finishes.
- Provide custom detailing for bases and trims.
- Specialty fixtures and casework

Equipment

- Lavatory, water closet, and combination bath and shower

SARAH'S BEDROOM

Planning

- Sarah will use this bedroom exclusively when she reaches age 5 or 6.
- Plan for sleeping comfort and personal clothing storage and toy storage.
- Storage should be easily accessible for the children.
- Provide a safe way for Sarah to escape in case of fire.

Furniture

- Full- or double-size bed with related furnishings
- Low table for miniature dollhouse
- Any furnishings that you find appropriate
- Plants, 911 light, accessories, and artwork

Design

- Carpet floor covering is required.
- Provide custom detailing for bases and trims.
- Provide custom bedding and window treatment.

Equipment

- Telephone
- House intercom panel
- Future computer hookup

JOHN'S BEDROOM

Planning

- John will use this bedroom exclusively when he reaches age 4 or 5.
- Plan for his sleeping comfort, clothing storage, and toy storage.
- Storage should be easily accessible for the children.
- Provide a safe way for John to escape in case of fire.

Furniture

- Full- or double-size bed with related furnishings
- Low table for train sets or car race tracks
- Any furnishings that you find appropriate
- Plants, 911 light, accessories, and artwork

Design

- Carpet floor covering is required.
- Provide custom detailing for bases and trims.
- Provide custom bedding and window treatment.

Equipment

- Telephone
- House intercom panel
- Future computer hookup

CHILDREN'S BATHROOM

Planning

- Sarah and John will use this bathroom.
- Provide storage for linens and their grooming items.

Furniture

- Child's step stool
- Accessories, artwork, and area carpets

Design

- Nonskid hard surface floors and walls are required.
- Provide selection of wall and ceiling finishes.
- Specialty fixtures and casework
- Window treatments are required.

Equipment

- Lavatory, water closet, and combination bath and shower

LINEN ROOM

Planning

- Provide storage for linens and a cedar storage area for clothing.
- Provide an ironing center work area that remains available 24/7 for quick use.

Design

- Provide selection of wall and ceiling finishes.
- Provide shelving and/or cabinet storage, bases, and trims.
- Hard surface flooring is required.

LAUNDRY

Planning

- Locate a small laundry center on the second level. Area must be concealed when not in use.
- Plan for convenience and function, including necessary appliances, counters, and fixtures as space allows.

Design

- Provide selection of wall and ceiling finishes.
- Provide shelving and/or cabinet storage, bases, trims, and doors.

- Hard surface flooring is required.

Equipment

- Washing machine, clothes dryer, and laundry sink
- Telephone

PLAYROOM AND EXERCISE ROOM

Planning

- Plan an exercise area with equipment and an aerobics area with a television.
- Children's play area should include a worktable and chairs and ample storage for games and toys.
- The family must be able to watch television from the exercise and play area.
- Allow sufficient space for child's free play.

Figure 25–10 Playroom and Exercise Room

Furniture

- Any furnishings appropriate for the space
- A place to sit down

Design

- Provide selection of wall and ceiling finishes.
- Provide shelving and/or cabinet storage for weights, videos, DVDs, toys, games, and craft supplies.
- Provide selection of bases and trims.
- Hard surface flooring is required.
- Floor impact mats for exercise and aerobic area are required.

Equipment

- House intercom panel
- Exercise equipment, including a rowing machine, ski machine, treadmill, universal weight machine, weight bench, pull-up bar, and free weights (storage)
- Television, VCR, DVD, and future computer equipment
- Portable telephone

Lower Level Spatial Requirements

CLUB ROOM

Planning

- Raul and his guests will use this room primarily.
- Plan for a fully functioning bar area that seats two.
- Plan a gaming table area and chairs for four people.
- Provide a secondary seating arrangement of lounge chairs adjacent to the game table with television viewing.
- Plan a seating area for conversation or watching television for four to six people.
- Plan a wine-tasting area with 4 to 6 linear feet of open cabinet storage for fine wines and a table with stools for four people. Incorporate full-height storage wall units, including shelving and cabinets to store wine, antique books, and golf-related awards and accessories.

Furniture

- Sofas, chairs, tables, lamps, bar stools, accessories, and artwork
- Cigar humidor

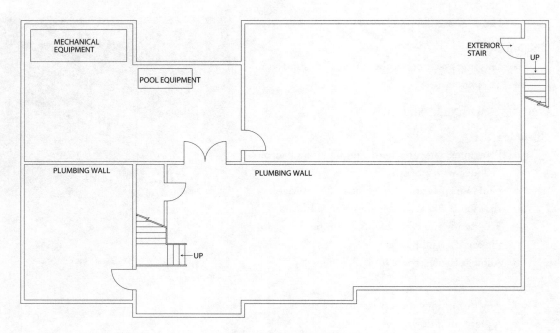

Figure 25–11   Lower Level Floor Plan

- Gaming table
- Combination backgammon and checkers table with two chairs

Design

- Carpet floor covering is required.
- Provide selection of wall and ceiling finishes.
- Provide custom detailing for cabinets, bar, bases, and trims.
- Provide granite countertops at the bar.
- Wood wall and ceiling surface treatments are required.

Equipment

- Large, flat-screen 42″ television and VCR and DVD equipment
- Bar: Dishwasher Sink, wine cooler, refrigerator, and ice machine
- Electronic dart board
- Telephone
- House intercom panel

BATHROOM

Planning

- This is the only bathroom for this level.
- Provide storage for linens and grooming items.

Furniture

- Bath accessories, door hooks, artwork, and area carpets

Design

- Hard surface floors and walls are required.
- Provide selection of wall and ceiling finishes.
- Provide custom detailing for bases and trims.
- Provide general illumination and task lighting for the vanity, water closet, and shower.
- Provide custom cabinet detailing and granite countertops.
- Specialty fixtures and casework

Equipment

- Lavatory, water closet, and enclosed shower

LAUNDRY/SEWING CENTER

Planning

- Plan the laundry for convenience and function, including necessary appliances, counters, and fixtures. Provide an ironing center work area.
- Plan a sewing center area complete with computer work surface, sewing work surface, counters, cutting worktable, and storage.

Furniture

- Layout and cutting table
- Sewing task chair
- Counter stools
- Built-in counters and storage units, including shelving and drawers
- Waste receptacles, hanging rods, tack board, and lamps

Design

- Provide selection of wall and ceiling finishes.
- Provide shelving and/or cabinet storage, bases, and trims.
- Provide general illumination and adequate task lighting for ironing and sewing and cutting area.
- Hard surface flooring is required.

Equipment

- Washing machine, clothes dryer, and laundry sink
- Sewing machine
- Computer (with pattern program) and wide-carriage printer
- Telephone
- House intercom panel

BAKING KITCHEN

Planning

- This kitchen is used primarily for baking. Caterers use it during large parties. Develop a functional and convenient arrangement with these functions in mind.
- Provide a deep double stainless steel sink and dishwasher area with counter space on each side. Food disposal will be located in the center sink.
- Provide a combination refrigerator and freezer and adjacent counter area.
- Provide a white marble baking counter work area with a built-in mixing center.
- Provide a small cooktop and two wall ovens, and proper ventilation.
- Provide wall and base cabinets for storage of cookbooks, utensils, bowls, freezer ware, baking appliances, cookware, and bake ware. Incorporate suitable cabinet features.
- Provide shelving in the pantry area for food and appliance storage.

Furniture

- Two counter stools

Design

- Provide selection of wall and ceiling finishes.
- Hard surface flooring is required.
- Provide selection of bases and trims.

Equipment

- Refrigerator, dishwasher, cooktop, hood, two ovens, and a microwave
- Double sink and disposal
- Telephone
- Fire extinguisher
- House intercom panel

STORAGE ROOM

Planning

- This room will store luggage, seasonal furniture and decorations, and some garden and pool maintenance tools.
- The pool equipment is installed here and must remain accessible for maintenance and repairs.

Design

- Provide selection of wall and ceiling finishes.
- Provide hard surface flooring.
- Provide selection of bases and trims.
- Provide built-in shelving suitable for heavy items, minimum 18″ deep.

SHELTER ROOM

Planning

- In preparation for emergencies they have created an emergency communications plan, enrolled in a first aid and AED/CPR course, and assembled a disaster supplies kit. As recommended by the American Red Cross it "includes special needs items for each member of the household (infant formula or items for people with disabilities or older people), first aid supplies (including prescription medications), a change of clothing for each household member, a sleeping bag or bedroll for each, a battery powered radio or television and extra batteries, goggles, sealing tape, food, bottled water and tools. It also includes cash and copies of important family documents (birth certificates, passports and licenses)."
- Additionally, they have designated a predetermined meeting place away from the home should their home be affected by an emergency or the area evacuated by authorities.
- In the event that the authorities or the weather conditions require them to "shelter in place"[2] for protection, this room will be the designated spot.
- The Van Patens have asked that you include furniture to address physical and psychological needs, providing utility, solace, rest, and privacy.

Equipment

- Telephone and cell phone
- Emergency generator
- House intercom panel

---

2. American Red Cross.

SUBMITTAL REQUIREMENTS

Phase I: Programming   Part A Due: _____   Part B Due: _____

A. Research local authorities, perhaps an insurance agent and all resources for safe rooms and shelter rooms. Using brainstorming techniques, fully explore the topics as they apply to the design of interior spaces.

B. Provide a one-page typed concept statement of your proposed planning direction and initial design concept.

Presentation formats for Parts A and B: White bond paper, 8½″ × 11″. Insert in a binder or folder.

Evaluation Document: Programming Evaluation Form

Phase II: Conceptual Design   Part A Due: _____   Part B Due: _____

A. Develop preliminary plans based on the functional and aesthetic requirements necessary, including use of space, furniture layouts, and fixtures.

B. Prepare preliminary furniture budget based on the above plans.

Evaluation Document: Conceptual Design Evaluation Form

Phase III: Design Development   Date Due: _____

A. Based on the approved preliminary plans, develop and present for review and approval final interior design recommendations to establish and describe the character of the interior space of the project.

B. Prepare documents and illustrations in the form of plans, elevations, four to six color and material sample boards, photographs of furnishings and fixtures, and four color renderings to show:

  a. Wall, furniture, and fixture arrangement
  b. Floor, wall, window, and ceiling treatments
  c. Furnishings, fixtures, and millwork
  d. Schematic lighting plan and lighting fixtures
  e. Colors, materials, and finishes
  f. Suggestions for art, accessories, and plants

Evaluation Document: Design Development Evaluation For

Phase IV: Final Design Presentation   Date Due: _____

The final presentation will be an oral presentation to a panel jury. Project must be submitted on 20″ × 30″ illustration boards. Your name, the date, the project name and number, the instructor's name, and the course title must be printed on the back of each board in the lower left-hand corner. Submit as many boards as required, including renderings, elevations, fixture and furniture selections, material selections, and plans.

The plans should be presented in 1/4″ scale. The elevations are to be presented in 1/2″ scale.

Evaluation Document: Final Design Presentation Evaluation Form

Project Time Management Schedule: Schedule of Activities, Chapter 31.

The AutoCAD LT 2002 drawing file name is vanpaten.dwg and can be found on the CD.

REFERENCE

Books

Designing Interiors. Interior Design Illustrated. Interior Design, Second Edition. Inside Today's Home, Sixth Edition. Total Design, Contemplate, Cleanse, Clarify, and Create Your Personal Spaces. Time-Saver Standards For Interior Design and Space Planning, Second Edition. Human Dimension and Interior Space: A Source Book of Design Reference Standards.

Product Manufacturers

American Saferoom Door Company: http://www.saferoom.com

André Julien: http://www.cuisines-aj.com

Ann Sacks: http://www.annsacks.com

Armstrong: http://www.armstrong.com

Jurs Architectural Glass: http://www.art-glass-doors.com

Octopus Pruducts, Ltd.: http://www.octopusproducts.com

Permagrain Products, Inc.: http://www.permagrain.com

Snaidero Kitchens: http://www.snaidero-usa.com

Project References

American Red Cross: http://www.redcross.org

Anser Institute for Homeland Security: http://www.homelandsecurity.org

READY.GOV: http://ready.gov/

Security World International: http://www.securityworld.com

U.S. Department of Homeland Security: http://www.dhs.gov

HISTORICAL FACILITIES

CORRIDOR

STAIR B

UP

4

PARLOR D

PARLOR C

3

TO THE GOLDEN STAIRS
TO CECILIAN HALL

CORRIDOR

LOBBY

VESTIBULE

ENTRY PORCH

DOWN

CORRIDOR

PARLOR B

PARLOR A

2

1

DOWN

DOWN

CHAPTER 26

Object D'Arte

OVTCOME SVMMARY

This project features an aesthetic aspect of the design process, with particular attention to decorative objects, which are often an integral part of your overall design concept. There are three phases to this featured project: identifying an object and doing some research on its history, the artist, and the process involved in the particular medium; analyzing and refining the research; and discussing the object in a presentation that both educates and engages the client.

client profile

For this project, your client has no restrictions regarding budget so as to enable you to select an object that you enjoy visually. Although a person of some wealth, your client is not well versed in the history of the decorative arts and is looking to you as a mentor to guide the acquisition of appropriate "things" to adorn either home or office.

project details

Because you have this financial freedom as a designer, select something that you absolutely love to look at for its visual qualities. You could even create an entire room for your client to enhance the object's presentation. Your Object D'Arte could be a historically significant piece that you have seen or studied; it could include a work by Tiffany, a chair by Renie MacIntosh, a table by Gustave Stickley, or an object created by a living artisan. You have free rein.

You may consult some arts, design, and craft journals, as well as some museum Web sites, to take a visual survey of the selection available to you.

SUBMITTAL REQUIREMENTS

Phase I: Research   Part A Due: _____   Part B Due: _____

    A. Identify the object completely. As you explore your work, take time to look, think, wonder, and ask questions. The initial concept is to get you (and your client) involved and respond to what you see. Journalists, storytellers, and visual historians include the five Ws when looking at something. Some questions to get you started follow:

- WHAT: What is it? What materials are used? What is the process? What are the colors? What are its dimensions? What is its function?
- WHEN: When was it created?
- WHERE: Where was it made? Where would this be best placed?
- WHO: Who made it?
- WHY: Why do you think the artist chose this form with these materials to complete the work?
- HOW: How does this object compare to others? How does this tell us about the person who created it? How does this reflect what we know about its time period?

    Write up your initial visual analysis.

    B. Research your object and present an annotated bibliography of your sources. Look for factual information on the artist, maker, producer, patron, process, historical period, etc.

Phase II: Analysis   Date Due: _____

    A. Refine your drafts from Phase I and organize your research with your visual analysis in a logical manner. Consider how you might introduce and describe your object.

    B. Attain images of the object being presented: for three-dimensional items, several views are desirable.

    C. Submit your research in final copy; include a bibliography.

Phase III: Presentation

    The final is an oral presentation to a panel or client. You must present images of your object along with information from Phase I and Phase II in order to engage your client with the piece so that she or he will ultimately purchase it. You should attempt to involve the panel or client while you educate by asking questions; consider this a discussion rather than a "lecture."

NOTES

CHAPTER 27

The Parlors, Seton Hill University

OUTCOME SUMMARY

This project features three phases of the design process: the designer will renovate an existing historical space by identifying the client's needs through research and analysis, by developing an initial design concept, and by drafting the documents in preparation for the final bidding for the project.

The designer will attend to the unique features presented by four rooms, coupled around a grand, second floor reception area, located in a turn-of-the-century academic building. Designers will be faced with constraints of historic appropriateness, as well as inclusion of existing objects and aesthetic elements, for example, fireplaces, hand-painted stenciling, moldings, and the incorporation of some existing furniture while being fiscally responsible with new purchases.

client profile

In the nineteenth century, the Sisters of Charity founded what was to become Seton Hill University. Initially, it was an academy for young women, which soon developed into a college; the first degree was awarded in music. The mission of the Sisters was guided by the philosophy of their founder, Elizabeth Ann Seton, who advocated that her followers be educated: "I am preparing you for the world in which you will live."

For most of the twentieth century, Seton Hill was recognized as an outstanding Catholic liberal arts college for women. Its mission to educate has been greatly expanded to include males, as well as a rich representation of international students. As the college's student population grew, so did the academic programs and the degrees offered; it is now a university that remains connected to its rich heritage and traditions while continuing to prepare all students for the ever-changing world in which they will live.

Likewise, the physical campus of Seton Hill University is architecturally diverse. The styles include Norman Revival, Late Victorian, and Modern, with plans for a Postmodern complex construction. Like many campuses, it is visually eclectic; one can date the "boom" years of growth

Figure 27–1
Exterior view of the Administration building

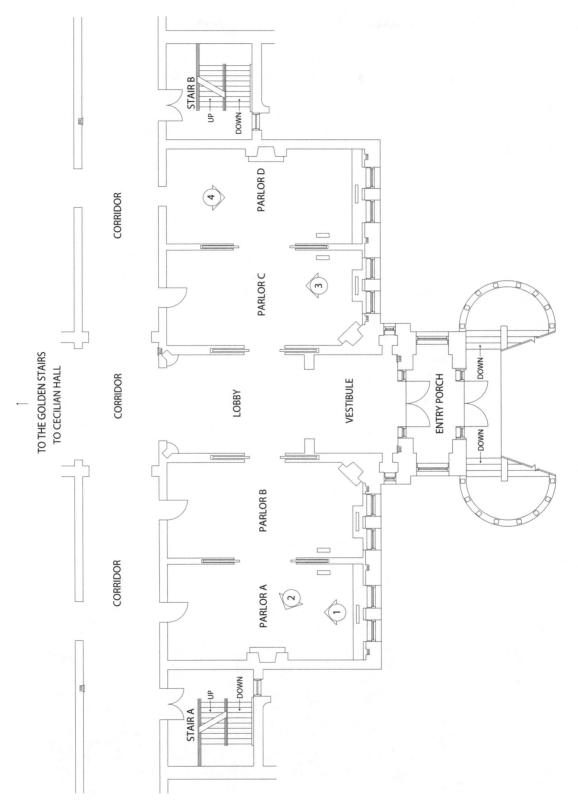

Figure 27-2 Floor plan with photograph annotations

by taking a walk on campus to see when the institution physically expanded. Interestingly, one of the modern residence halls was designed by Philip Johnson in the International style. In the early 1990s, Johnson drew up plans for a Fine Arts (Visual and Performing) structure, evoking a contemporary expression of a Tuscan hill town, dominated by a campanile.

Seton Hill, as its name implies, is situated on a hill overlooking Greensburg, Pennsylvania. The administration building, while not the oldest site on campus, is the most physically dominant as well as the historic heart and soul of the institution. In this building reside several vital places. The spirit of Administration is St. Joseph's Chapel where liturgy opens and closes each academic year; many graduates choose to be married here. The heart and its arteries run throughout the structure. The upper floors have housed novices as well as student dormitories and now serve students in the academic support programs. Cecilian Hall, located on the second floor, hosts concerts, honors convocations, and visiting scholar lectures.

Also located on the second floor, the Parlors are gracious old rooms where students are welcomed into their freshman year by the faculty in a formal receiving line; the Parlors are the site of a variety of academic as well as social receptions. Thus, administration, and the Parlors in particular, hold a unique place in the lives and memories of each Setonian.

project details

General and Architectural Information

Location: Seton Hill University, Administration Building, Greensburg, PA 15601

Occupancy: Assembly

Gross square footage: 2,400. Each Parlor is approximately 600 square feet.

Facilities Required

Space: Each of the four Parlors measures 31′ 0″ × 15′ 4″, and is flanked by a grand hallway that leads into Cecilian Hall on the same floor, and up a carved oak staircase to St. Joseph's Chapel. (The Sisters refer to this as "The Golden Stairs.") There are pocket doors connecting each set of rooms; this open space measures 9′ 8″ when opened. The ceilings are high and measure 14′ 0″. Each room has two shuttered windows that are 4′ 0″ wide and 10′ 6″ high.

Figure 27–3 Parlor A. Photo 1.
All color photographs are located on the CD

Figure 27–4 Parlor A. Detail of ceiling

Figure 27–5
Parlor A. Photo 2: Fireplace

Figure 27–6   Parlor A. Detail of fireplace

Figure 27–7
Parlor C. Photo 3

Figure 27–8   Parlor C. Detail of ceiling

Figure 27–9   Parlor D. Photo 4

Figure 27–10   Parlor D. Detail of ceiling

SUBMITTAL REQUIREMENTS

Phase I Research and Analysis   Date Due: _____

    A. Research the historical period of the administration building.

    B. Prepare two boards of adjoining Parlors, including furniture used and spatial arrangements.

    C. Prepare a one-page typed summary of the concept statement for your proposed plans.

Phase II: Conceptual Design

Part A Due: _____   Part B Due: _____   Part C Due: _____

    A. Develop plans from Phase I into a final floor plan for all four Parlors. It is to be completed using the CADD program.

    B. Develop colors and materials schemes, including pictures.

    C. Develop perspective drawings of the interiors.

Phase III: Final Design   Date Due: _____

The final presentation is to be an oral presentation to a panel jury. The project must be submitted on 20″ × 30″ illustration boards. Your name, the date, the project name and number, the instructor's name, and the course title must be printed on the back of each board in the lower left-hand corner. Submit as many boards as required, including a colored rendering of each room, furniture selections, plans, etc. Plans should be presented in 1/8″ scale.

Evaluation Document: Final Design Presentation Evaluation Form

Project Time Management Schedule: Schedule of Activities, Chapter 31.

The AutoCAD LT 2002 drawing file name is parlors.dwg and can be found on the CD.

MEDICAL FACILITIES

EXIT ELEC PANEL

UP

LAV.

MECH.

ENTRANCE

CHAPTER 28

Office of Dr. Gene Cordova

OUTCOME SUMMARY

The project features three phases of the design process: identifying and analyzing the client's needs and goals, developing conceptual skills through schematic or initial design concepts, and detailing and refining ideas from the schematic design phase. This assignment focuses on increasing knowledge of planning and design for medical offices, in particular an optometry practice that includes eyewear dispensing.

client profile

Dr. Cordova is an optometrist with a thriving private practice that includes eyeglass-dispensing services. He is relocating his offices to another facility and requires interior planning and design services. After a rather friendly and warm greeting from him, the first thing you will notice is a kind demeanor, and the second is his exceptional height. He has been in practice for over 20 years, teamed with his wife, who assists with the dispensing services. He sees at least three patients per hour except on Wednesdays, of course, when he hits the links for a minimum of thirty-six holes of golf and sometimes more. Not surprising that he is a scratch golfer. However, consider yourself very fortunate to be invited to dine at his home: his hobby is cooking and it outshines his golf game.

He has a hectic exam schedule. The majority of his patients are elderly and the demand for eye care has increased with the growing elderly population in addition to vision problems resulting from the use of computers. An examination lasts approximately 30 to 60 minutes, depending on the procedures required for the patient exam. The patient will check in at the reception, wait to be called, taken to the Visual Fields testing area, moves next to an exam room for a glaucoma test and/or to an exam room to be checked by the doctor. The patient will then be fitted for eyeglass frames or trained for contact lenses. If eye drops are given that temporarily blur vision, the patient is taken to the waiting area for twenty minutes and given disposable protective glasses before leaving. The visit is completed at the reception station where the bill is paid and appointments are made for the next visit of fitting the eyeglasses with corrective lenses.

Typically, an inquiry is performed regarding a new patient's complete medical history before any exam begins. This is documented on a form that the patient completes in the waiting area. The examination includes an inspection of both the interior and exterior of the eye to look for eye or systemic disease. Tests to measure the ability to see clearly and sharply at any distance are performed. Measurement is taken of the eyes' ability to focus light rays, eye muscle control and coordination is checked, testing focus abilities, and a glaucoma test is performed. Other special tests performed may be field of vision, depth perception, and color perception as required. Exams may result in a treatment plan that includes corrective lenses or vision therapy and, in some cases, prescriptions for medication or referral to a specialist.

The dispensing services include the selection of eyeglass frames, lenses, and lens coatings. Once the determination is made for the eyewear, a work order is prepared and sent to the ophthalmic laboratory with the information needed to grind and insert the lenses into the frames. Back at the lab, it is verified that the lenses have been ground to the prescription specifications. The next step is to fit, educate the patient on the use and care, and also for new corrective lens wearers, there is an instruction on adapting to the eyewear. Contact lenses are also dispensed. Fitting this type of lens requires skill, care, and patience. Special instruments and microscopes are used to observe the eye in this process. Instruction is given on insertion, removal, care, and assurance of fit.

The office will be located in a single-story, small strip center with two other tenant spaces. Its location is visually accessible to the surrounding neighborhood and can be seen from adjacent street intersections. The site includes customer parking in front and employee parking at the rear entry of the space. The building is finished with gray brick and aluminum window and door storefront frames. The interior will be finished with gypsum board walls, and you are required to use the existing $2' \times 4'$ ceiling grid ($8'\ 10''$ above the finished floor) and the existing $2' \times 4'$ fluorescent light fixtures. The light fixtures are recessed type and you may relocate the fixtures, supplement the fixtures, and change the lens. Interior windows are 33 inches above the floor and 69 inches high. Dr. Cordova has requested a simple metal blind window covering solution to block the heat from the sunlight when required.

project details

General and Architectural Information

Location: 155 Pike Street, Boston, Massachusetts

Gross square footage: 1,800. Net square footage: 1,645

Initial number of employees: Four. Projected number of employees: Six in 1 year

Existing construction. Floors are concrete, ceilings are $2' \times 4'$ lay-in tile and the walls are drywall. Architectural plans are not available. Site information gathering and survey is required. The architect/contractor will administer construction documents and the bidding of the interior furniture, fixture, and finish work.

Consultants required: Architect, electrical engineer, and mechanical engineer

Codes and Governing Regulations

Occupancy: B Business

Building Type: Noncombustible single-story structure. Gross building area: 6000 square feet

Fire Protection: No fire protection or suppression system

Follow ADA guidelines and the specific Life Safety and Building Codes adopted by the your local government for this exercise. Finishes must meet code requirements, be easily maintained, and be durable.

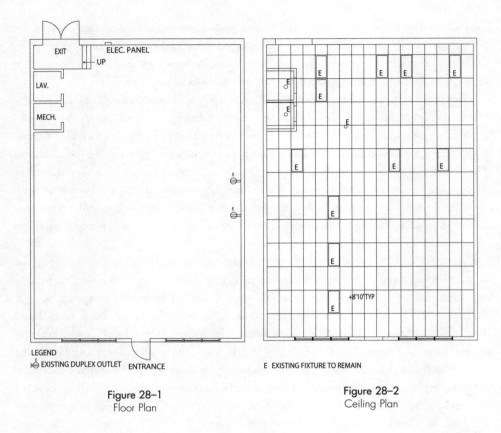

LEGEND
E EXISTING DUPLEX OUTLET   ENTRANCE

E EXISTING FIXTURE TO REMAIN

Figure 28–1
Floor Plan

Figure 28–2
Ceiling Plan

Facilities Required

Reception, Business Office, Waiting Room, Showroom/Fitting Area, Lab, Delivery and Adjustment, Contact Lens Training Area, Visual Fields, Exam Room 1, Exam Room 2, Exam Room 3, Consultation Room, and Storage. (The restroom exists and is not to be altered.)

RECEPTION

Planning

- Adjacency: Entry Door, Waiting Room, and Business Office. Locate as part of the Business Office. 60 square feet.
- The receptionist will greet patients who are checking in for appointments, arrange appointments, answer the telephone, have secretarial support duties, gather patient information, and receive payments.
- Allow for adequate queuing space, circulation space, and create a traffic passageway to the waiting area and Exam rooms.
- Complete vision of the patient Waiting Room and Showroom/Fitting Area is required as well as security and confidentiality for the office.
- Plan a counter configuration that has transaction surfaces for patients and work surfaces for the receptionist and/or clerical staff.

Furniture

- Custom or modular[1] counter and work surface
- Task chair, file drawer, pencil drawer, box drawer, and retractable keyboard tray for the receptionist

---

1. Modular refers to a furniture system or the equivalent.

- Area carpets, waste receptacle, clipboards, pen and pencil holders, appointment cardholders, and visuals, if appropriate

Design

- Provide selection of wall and ceiling finishes.
- Provide Reception and transaction counter design and finishes.
- Hard surface flooring is required.

Equipment

- Telephone
- Computer

BUSINESS OFFICE

Planning

- Adjacency: Reception and Waiting Room. It can be contiguous with the receptionist work area. 180 square feet.
- Provide a work area for a bookkeeper and a billing clerk.
- Provide an area for patient records and general file storage.
- Include a work area for office equipment and related supplies.

Furniture

- Work surfaces, task chairs, file storage, box drawers, pencil drawers, retractable keyboard trays, and task chairs
- Patient medical record storage units of 60 linear feet in an open-file system for side tab folders
- 24 linear feet of general lateral file storage

Design

- Provide selection of wall and ceiling finishes.
- Carpet floor covering is required.
- Provide general illumination and task lighting.

Equipment

- Computer and telephone at each station; computers are networked
- Copy machine
- Fax machine
- Mail postage equipment

WAITING ROOM

Planning

- Adjacency: Reception. 140 square feet
- This area is used for patients to wait until it is time for their eye exam. It is also the waiting area after receiving eye drops.
- Plan seating for eight people.

Furniture

- Combination of sofa groupings and chairs. Be mindful of seating comfort and function requirements for the elderly. Include space for patients in wheelchairs.
- Tables, reading lamps, accessories, artwork, and plants
- Magazine racks and brochure display for medical information
- Coat, hat, and umbrella storage

Design

- Create a pleasurable atmosphere with subdued lighting.

- Provide selection of wall and ceiling finishes.
- Carpet floor covering is required.

SHOWROOM/FITTING AREA

Planning

- Adjacency: Reception and Waiting Area. 275 square feet
- Patients will be able to select eyeglass frames in this area and be fitted for the frames.
- Plan for two fitting stations with a chair for the patient and a stool for the doctor on the drawer side of the station.
- Provide 19 linear feet of wall display of eyewear frames, and floor fixtures of a total of 8 linear feet of display.

Furniture

- Fitting stations ($2' \times 4'$) with drawers and storage
- Tabletop fitting mirrors
- One stool and one chair at each fitting station

Design

- Create definition and a good visual presentation for the frame display area. Eyeglass frame display wall with slat wall, continuous base cabinet and display lighting is required.
- Provide selection of wall and ceiling finishes.
- Carpet floor covering is required.

LAB

Planning

- Adjacency: Showroom/Fitting Area. 36 square feet
- Adjustments are made to corrective eyewear in this area.

Furniture

- Waste receptacle

Design

- Provide wall cabinets and counters with a 3″ backsplash.
- Provide shadow-free lighting.
- Provide selection of wall and ceiling finishes.
- Hard surface flooring is required.

Equipment

- Computer on network
- Frame warmer
- Lensometer
- Microscope
- Small round stainless steel sink
- Wall-mounted telephone
- Plug-in strip outlets are required at counter height. Provide a total of eight outlets.

DELIVERY AND ADJUSTMENT

Planning

- Adjacency: Lab and Showroom/Fitting Area. 70 square feet
- The functions of this area are to dispense correction eyewear, educate the patient on use and care of eyewear, test prescription of eyeglasses, and fit glasses.

Furniture

- Fitting station (2′ × 4′) with drawers and storage
- Tabletop fitting mirrors
- One stool and one chair

CONTACT LENS TRAINING AREA

Planning

- Adjacency: Exam Rooms and Visual Fields. 70 square feet
- The primary function is to dispense and train for the use of contact lens correction eyewear.
- Plan for two fitting areas where the patient sits on one side and the doctor on the other side.
- Provide a storage area and a counter with sink.

Furniture

- Mirrored-top fitting tables (2′ × 4′)
- Cabinets with drawers and 3″ backsplash at the countertop
- Two adjustable stools
- Waste receptacles

Design

- Provide selection of wall and ceiling finishes.
- Provide selection of finishes for cabinets.
- Carpet floor covering is required.
- Provide general illumination and specialty task lighting.

Equipment

- Two duplex outlets

VISUAL FIELDS

Planning

- Adjacency: Contact Lens Training Area. 100 square feet
- This is a multipurpose testing area that includes Visual Fields tangent screen testing and automated refraction and fundus exam.
- Tables to support equipment are required for the fundus and automated refraction test.
- Wall surface is required for the fields test.
- The aide performs the tests.

Furniture

- Two tables (2′ × 4′)
- Three adjustable stools

Design

- Provide selection of wall and ceiling finishes.
- Carpet floor covering is required.
- Dimmer controlled general lighting is required.

Equipment

- Automated refractor, tangent screen (1 meter)
- Fundus camera
- Ophthalmoscope
- Two duplex outlets

EXAM ROOM 1

Planning

- Adjacency: Waiting Room and Consultation Room. 84 square feet
- Primary exam functions are the projection eye chart refraction testing and slit lamp exam for the cornea, iris, and lens.
- Dr. Cordova will work off the instrument stand and the counter. Equipment and instruments are grouped around the patient examining chair, and Dr. Cordova sits at the right of the patient. The instrument stand is to be situated to the left of the patient chair.
- Plan a counter area with knee space, including a sink and storage for medications and miscellaneous instruments.
- A separate access door for the doctor is required (preference is a pocket door).

Furniture

- Chart holders to be located outside of the doctor's entry door
- A counter (5′ 0″ in length) with drawers and storage
- Five-leg stool with casters for the doctor
- Instrument table
- Two wall-mounted mirrors located in front of the patient chair
- Wall-mounted brackets for equipment
- Accessories and artwork
- Magazine racks and brochure display
- Waste receptacle

Design

- Provide selection of wall and ceiling finishes. (No patterns on walls.)
- Hard surface flooring is required.
- Only one general illumination fixture is required. Lighting on dimmers controlled from the wall at the door and counter is required.

Equipment

- Trial lens box sets
- Instrument stand
- Chart projector controlled from the counter switch
- Fixation light, 72″ high, wall-mounted behind the patient chair
- Slit lamp mounted on the instrument stand
- Wall-mounted screen located behind the patient chair
- Patient chair
- Wall-mounted telephone
- Power requirements: plug-in strip outlets are required at counter height, a total of four outlets. Quadruplex outlet required behind the patient chair; power outlet for the examining chair
- Small round stainless steel sink

EXAM ROOM 2

Planning

- Adjacency: Waiting Room and Consultation Room. 84 square feet
- Same as Exam Room 1.

Furniture

- Same as Exam Room 1.

Design

- Same as Exam Room 1.

Equipment

- Same as Exam Room 1.

EXAM ROOM 3

Planning

- Adjacency: Waiting Room and Consultation Room. 100 square feet
- This room is used as exam room, for emergency visits, and as a data collection room. Glaucoma testing and tangent screen testing and other procedures as required are performed here.
- Same as Exam Room 1.

Furniture

- Same as Exam Room 1.

Design

- Same as Exam Room 1.

Equipment

- Same as Exam Room 1.

CONSULTATION ROOM

Planning

- Adjacency: Exam Rooms. 84 square feet
- Dr. Cordova will meet with patients here infrequently for special consultation. It is his study and research area, a place to review business matters, receive and make calls, and rest.
- This room may become future Exam Room 4.
- The room is very small. You may elect to use modular furniture in any configuration equal to that described below.

Furniture

- Executive desk, chair, credenza, book storage, literature storage, and file storage
- Side chairs
- Accessories; frames for degrees, licenses, and certificates; and artwork
- Waste receptacles, task lighting, and desk accessories

Design

- Provide selection of wall and ceiling finishes.
- Carpet floor covering is required.

Equipment

- Computer (on network)
- Telephone

STORAGE

Planning

- Plan 20 square feet of storage.
- Shelving is required for office supplies, cleaning supplies, and miscellaneous storage.

Design

- Provide selection of wall and ceiling finishes.
- Hard surface flooring is required.

SUBMITTAL REQUIREMENTS

Phase I: Programming   Part A Due: _____   Part B Due: _____

Part C Due: _____   Part D Due: _____

   A. Research various codes, arrangements and clearances for the above-outlines spaces. Research fixture, furniture, and equipment for the same.

   B. Analyze the program outlined above and develop a programming matrix keyed for adjacency locations. A sample blank matrix is provided in Chapter 31.

   C. With the information gathered from the matrix, provide a minimum of three bubble diagrams for the spatial relationships of the areas required.

   D. Provide a one-page typed concept statement of your proposed planning direction and initial design concept.

   Presentation formats for Parts A through D: White bond paper, 8½″ × 11″, with a border. Insert in a binder or folder. The matrix and bubble diagrams are to be completed using the CADD program.

   Evaluation Document: Programming Evaluation Form

Phase II: Conceptual Design

Part A Due: _____   Part B Due: _____

   A. Develop a preliminary partition and furniture plan, showing all walls, ceilings, built-in fixtures, doors, and windows. These plans can be freehand drawings or CADD drawings in 1/4″ scale. Drawings must have an appropriate border. Finished drawings must be a blueprint or white bond copy.

   B. Develop the above plans into a final floor plan, showing all items noted in the program. This plan is to be completed using the CADD program. Develop ceiling and lighting studies into a final reflected ceiling plan, showing all ceiling conditions and heights using the CADD program.

   Evaluation Document: Conceptual Design Evaluation Form

Phase III: Design Development

Part A Due: _____   Part B Due: _____

   A. Develop a color, material, and final furniture scheme. Samples and pictures can be presented in a loose format.

   B. Develop a perspective sketch of the interior of the Showroom/Fitting Area, showing all design elements and furnishings.

   Evaluation Document: Design Development Evaluation Form

Phase IV: Final Design Presentation   Date Due: _____

   A. The final presentation will be an oral presentation to a panel jury. Project must be submitted on 20″ × 30″ illustration boards. Your name, the date, the project name and number, the instructor's name, and the course title must be printed on the back of each board in the lower left-hand corner. Submit as many boards as required, including a colored rendering, fixture and furniture selections, material selections, and plans.

B. The plans should be presented in 1/4″ scale. The elevations are to be presented in 1/2″ scale.

Evaluation Document: Final Design Presentation Evaluation Form

Project Time Management Schedule: Schedule of Activities, Chapter 31.

The AutoCAD LT 2002 drawing file name is cordova.dwg and can be found on the CD.

REFERENCES

Books

Malkin, Jain. 2002. *Medical and dental space planning.* 3rd ed. New York: Wiley.

Piotrowski, Christine M., and Elizabeth A. Rogers. 1998. *Designing commercial interiors.* New York: Wiley.

Equipment Manufacturers

C.I.O.M. Optical Instruments: http://www.studio-q.com/ciom/english/azienda.html

Frastema (including chairs): http://www.frastema.com

Mainline Optical, Ltd.: http://www.main-line.co.uk.

Nikon Optical Equipment: http://www.nikon.com

Opticians Association of America (links to equipment): http://www.oaa.org

Tomey Corporation USA: http://www.tomey.com

NOTES

PERSONAL SERVICE FACILITIES

OPEN KITCHEN

REST ROOMS

RESTAURANT

LIGHT WELL
OPEN TO EXTERIOR

SALON AREA

CHAPTER 29

Joie Salon

OUTCOME SUMMARY

This project is an exercise that covers the design process, from planning and design, through design presentation. The size of the space is small and the attention to small details is higher than expected. The preliminary submission requirements are in stages and the final presentation requirements are extensive, including renderings and budgets.

client profile

The owner of Joie Hair Salon is an award-winning stylist. He employs a staff of six full-time professional stylists, twelve part-time assistants, and two full-time front desk attendants. He has been a salon owner for over 15 years and is relocating his salon to a nearby building. He has decided to stay in the immediate location of his current salon to ensure making this transition easy for his clients and staff.

Hair services include cutting, washing, coloring, and conditioning. Manicures and pedicures are also available. Each stylist has two chairs. One chair is for cutting their client's hair and the other is for the next appointment. The assistant assigned to the stylist prepares the next client and the stylist's tools while the stylist is finishing a cut. Make-up services are provided for special events, such as weddings, galas, pageants, etc.

The upscale neighborhood is reminiscent of Georgetown, but on a smaller, more personal scale. Since the middle of the last century, boutique shops, restaurants, and bars have been family owned. Historically, they have been single station points of sale, but in the 1990s, the smaller unique storefronts slowly began to be replaced with scaled-down versions of the large chain stores typically found in enclosed shopping malls. There are over 100 independently owned businesses and approximately 20 nationally owned businesses, including professional services, specialty fashion boutiques, upscale furniture stores, galleries, salons, restaurants, and clubs. You can also find some funky resale shops such as Hey Betty!, famous for their high-quality, vintage clothing. The community sponsors events such as Jam on Walnut, a music festival featuring local talent; the Art Festival on Walnut Street, featuring artisans and craftspeople from around the world; the City of Pittsburgh marathon; and the Vintage Grand Prix Car Show.

The new location is on the top floor of a three-story building. The Class-A building was completely renovated in the 1980s at the front end of the latest building renaissance. The building is on the most prominent corner in the neighborhood. The entire third floor was a restaurant for the previous five years. The architectural approach evokes the mid-1980s style of Morphosis Architects made significant by Thom Mayne through projects such as Kate Mantilini Restaurant in Beverly Hills, California, and 72 Market Street Restaurant in Venice, California.

The salon needs approximately 1,400 square feet of the 4,000 available; therefore, the floor is being subdivided for the salon. This will leave space for another tenant on the north side of the floor. There are windows on the south and west facades of the tenant space. Since the salon was completed, a restaurant with an Asian theme leased and moved into the adjacent space. The building provides separate toilet facilities for patrons and employees.

The owner requests that the new space take advantage of the perimeter windows for the benefit of both clients and stylists. The former location of the salon was on the same street; however, the space was dark and not very inviting. Style, fashion, and music are concepts that relate nicely, and this location is no exception. Create an environment that is forward-looking, thematic, exciting to the senses, and, of course, memorable. You are required to select all wall, floor, ceiling materials, and finishes, and appropriate fixtures and furniture. A practical request is to use finish materials that are durable and easy to maintain. Hair clippings become easily lodged in soft surfaces and hair colors stain most materials quickly. And finally, plan the lighting and select lamp sources to complement the client's aesthetics.

project details

General and Architectural Information

Location: Joie Hair Salon, 5401 Walnut Street, Pittsburgh, Pennsylvania

Gross Square Footage: 1,400 (tenant space)

Initial number of employees: Twenty. Projected number of employees: Twenty within 1 year

Hours of operation: Monday–Friday: 9 a.m. to 9 p.m.; Saturday: 10 a.m. to 7 p.m.

Existing construction: Bidding is required. Facility plans are available. Site information gathering is required. Maximum ceiling height is 11′ 6″.

Consultants required: Architect, mechanical engineer and electrical engineer

Codes and Governing Regulations

Occupancy: Business

Building Type: The Pennsylvania Labor and Industry Department classified this building as Ordinary Type Construction with a D-O Occupancy Group.

The 1993 BOCA Building Code requires a minimum of two exits from the floor. The location of the tenant space creates a special condition, which requires that one of the exits from the salon space must lead to a fire stair within the adjacent tenant space. Exit signs are required. A fire extinguisher is required. An automatic sprinkler system is required. You must comply with the basic provisions of standard regional building codes, fire and safety codes, and the Americans with Disabilities Act for retail spaces.

Because the building provides accessible toilet rooms on the floor, your only requirement for a toilet room is that the owner is requesting one within the salon for convenience. With this in mind, the toilet room (21 square feet) is not required to be fully accessible.

Facilities Required

Lobby, Waiting Area, Reception, Retail, Office, Changing Room, Styling Stations (six required), Manicure Stations (two required), Washing Stations (five required), Coloring Stations (four required), Mixing, and Utility, and Toilet Room

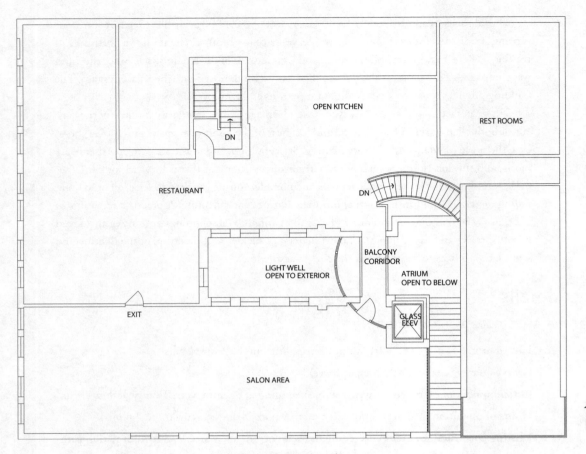

Figure 29–1   Floor Plan

Spatial Requirements

LOBBY

Planning

- Square footage: 85. Adjacency: Entry Door and Reception
- Make it convenient to achieve point of sale before client departure.

Furniture

- Furniture is not required but accessories, including an umbrella stand, and appropriate furnishings may be included.

Equipment

- Security alarm with keypad device

Design

- Space must reflect an instantaneous visual impact and set the theme for the shop.
- A logo is to be developed and incorporated into a sign area at the entrance.

WAITING AREA

Planning

- Square footage: 57. Adjacency: Lobby and Reception
- Provide comfort for short-term waiting.

Furniture

- Seating for four with related furniture
- Magazine storage

RECEPTION

Planning

- Square footage: 80. Adjacency: Retail, display, and Styling Stations
- Plan this space for two employees to work independently, including one salesperson and one appointment setter.
- This space is the primary access to Retail and to the display area and is the secondary access to the Styling Stations.

Furniture

- Check-in/out counter, approximately 15 linear feet, with product display (shampoo, conditioner, gels, nail polish, make-up, etc.)
- Storage for wrapping supplies and bags
- Two chairs with casters and two chair mats

Equipment

- Two personal computers (one for appointment scheduling and one with sales register drawer)
- Credit card machine
- Central telephone system

RETAIL

Planning

- Square footage: 16. Adjacency: Waiting Area and Reception
- Incorporate display for a variety of changing merchandise (jewelry, combs, brushes, sunglasses, etc.).

Furniture

- Approximately 30 linear feet of display cases and shelving

OFFICE

Planning

- Square footage: 37.5. Adjacency: Reception
- Provide a work area for the office manager.
- Provide storage for back stock of retail merchandise and products.

Furniture

- Work surface of approximately 30″ × 72″, task chair, file storage, box drawers, pencil drawers, retractable keyboard trays, locking base cabinets for storage, and guest chairs
- Waste receptacles, task lighting, bulletin board, and a marker board that is 36″ high × 48″ wide

Equipment

- Telephone
- Copy machine
- Fax machine

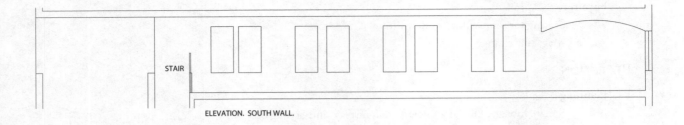

STAIR

ELEVATION. SOUTH WALL.

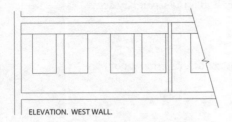

ELEVATION. WEST WALL.

Figure 29–2 Window Wall Elevation

CHANGING ROOM

Planning

- Square footage: 16. Adjacency: Near the entrance for immediate customer access
- Provide complete privacy for clients to change into cutting robe.
- Provide a small room which has garment and coat storage.

Furniture

- One chair, mirror, hamper, and purse shelf

Equipment

- Garment hooks
- Clothes rod with hangers for client clothes and shelf

SIX STYLING STATIONS

Planning

- Square footage: 413. Adjacency: Windows
- Open and airy, provide visual privacy between adjacent stylist space.
- Plan adequate space for four people (stylist, assistant, and two seated clients).
- All six stations should be identical, in case stylists switch locations.

Furniture

- Two salon chairs, each with a mirror
- An easily accessible work surface of approximately 30″ for the stylist
- Storage for styling equipment, accessories, and tools

Equipment

- Styling hand tools, blow dryers, curling irons, and electric cutters

MANICURE STATIONS (TWO REQUIRED)

Planning

- Square footage: 30
- Locate out of direct sight lines to provide privacy for clients.

- Provide space for two seated people, including the nail technician and the client.

Furniture

- Technician stool with casters
- Client chair
- Manicure table with casters
- Tool storage
- Product display shelves (approximately 6 linear feet)

Equipment

- Soaking trays for hands and for feet

WASHING STATIONS (FIVE REQUIRED)

Planning

- Square footage: 188.5. Adjacency: Coloring Stations
- Plan an open area, giving assistants client access from all sides.

Furniture

- Five chairs with reclining backs
- Shelving units for supplies and towels, total of twelve linear feet

Equipment

- Five hair washing sinks
- Televisions with cable

COLORING STATIONS (FOUR REQUIRED)

Planning

- Square footage: 87.5. Adjacency: Washing Stations
- Provide a sense of privacy.

Furniture

- Four chairs with swivel and reclining backs
- Floor-mounted base cabinets with countertop (approximately 14 linear feet)
- Wall-mounted cabinets (approximately 14 linear feet)
- Mirrors

Equipment

- Two heat towers to service four chairs

MIXING

Planning

- Square footage: 39. Adjacency: Coloring Stations and Utility
- This is a behind-the-scenes preparation area.
- It is generally used by one person at a time. Access is for staff only.

Furniture

- Floor-mounted base cabinets with countertop (approximately 6 linear feet)
- Wall-mounted cabinets (approximately 6 linear feet)

Equipment

- One utility sink in countertop

UTILITY

Planning

- Square footage: 45.38. Adjacency: Mixing
- This must be conveniently located for ready access by staff.
- Plan an enclosed room, which primarily functions as a laundry space.
- Provide storage for both dirty and clean towels and robes.

Furniture

- Floor-mounted base cabinets with countertop for folding
- Wall-mounted cabinets

Equipment

- Laundry tub
- Washing machine
- Clothes dryer

SUBMITTAL REQUIREMENTS

Phase I: Programming

Part A Due: _____   Part B Due: _____   Part C Due: _____

A. Research various codes, arrangements, and clearances for the above-outlined spaces. Research fixture, furniture, and equipment for the same. Analyze the program outlined above and develop a programming matrix keyed for adjacency locations. A sample blank matrix is provided in Chapter 31.

B. With the information gathered from the matrix, provide a minimum of three bubble diagrams for the spatial relationships of the areas required.

C. Provide a one-page typed concept statement of your proposed planning direction and initial design concept.

Presentation formats for Parts A through C: White bond paper, 8½″ × 11″, with a border. Insert in a binder or folder. The matrix and bubble diagrams are to be completed using the CADD program.

Evaluation Document: Programming Evaluation Form

Phase II: Conceptual Design   Part A Due: _____   Part B Due: _____

A. Develop a preliminary partition and area location plan, showing all walls, ceilings, built-in fixtures, doors, and windows. Label the required areas and rooms, and indicate the square footage required and the square footage planned. These plans can be freehand drawings or CADD drawings in 1/4″ scale. Drawings must have an appropriate border. Finished drawings must be a blueprint or white bond copy.

B. Develop the logo design and schematic elevations for the salon entry.

Evaluation Document: Conceptual Design Evaluation Form

Phase III: Design Development

Part A Due: _____ Part B Due: _____ Part C Due: _____

A. Develop the above plans into a final floor plan, showing all items noted in the program. This plan is to be completed using the CADD program. Develop ceiling and lighting studies into a final reflected ceiling plan, showing all ceiling conditions and heights on the CADD program.

B. Develop a color, material, and final furniture and fixture scheme. Samples and pictures can be presented in a loose format.

C. Develop a perspective sketch of the interior of the salon, showing all design elements and furnishings. Refine the front elevation of the salon and sections, showing all design and signage elements.

Evaluation Document: Design Development Evaluation Form

Phase IV: Final Design Presentation   Date Due: _____

A. The final presentation will be an oral presentation to a panel jury. Project must be submitted on 20″ × 30″ illustration boards. Your name, the date, the project name and number, the instructor's name, and the course title must be printed on the back of each board in the lower left-hand corner. Submit as many boards as required, including a colored rendering, fixture and furniture selections, material selections, and plans.

B Prepare a budget, including the cost of each furniture item selected. Be prepared to discuss the features of the finishes and furnishings chosen.

C. The plans should be presented in 1/4″ scale. The elevations are to be presented in 1/2″ scale.

Evaluation Document: Final Design Presentation Evaluation Form

Project Time Management Schedule: Schedule of Activities, Chapter 31.

The AutoCAD LT 2002 drawing file name is joie.dwg and can be found on the CD.

REFERENCES

Modern Salon Media: http://www.modernsalon.com
Worldhaironline.com: http://www.worldhaironline.com

NOTES

NOTES

SECTION
10

EXHIBIT
FACILITIES

CHAPTER 30

DI Exhibit

OUTCOME SUMMARY

The project features four phases of the design process: identifying and analyzing the client's goals, developing conceptual skills through schematic or initial design concepts, detailing and refining ideas from the schematic design phase, and building a scaled presentation model. The project focuses on developing design skills relating to exhibit spaces and model construction.

client profile

Danna Industries (DI), a mechanical and sheet metal contractor, plans to construct an exhibit to feature the company in construction industry trade shows. The primary goal of the exhibit is to identify the company and feature its capabilities and projects. The target audience will be architects, engineers, general contractors, facilities managers, and facility owners. Use your design skills to create a presentation that engages the visitor and leaves a lasting impression about the company. Your solution must incorporate the use of at least three principles of design.

The types of building specializations of DI are large retail facilities, educational institutions, hospitals and medical facilities, correctional institutions, commercial office buildings, sports complexes, airports, country clubs, theaters, hotels, conference centers, and performing art centers. The mechanical systems services provided are design-build, plan-spec, and design-assist HVAC installations. Also sheet metal ductwork production and mechanical piping systems design, fabrication, and installation are specialties of the company. Finally, DI commissions, services, controls, and maintains the systems.

The following are excerpts from the company brochure:

"Key areas of our goals include fair pricing, project planning, multi trade coordination, and most importantly 'On Time Construction'.

"With our computerized system of tracking purchase orders, shop drawings and labor; in addition to our labor and material cost controls, coordination and scheduling processes, we are confident in our ability to manage any project scope. The ability of our Project Managers to use our Estimating Department and their system to disseminate the piping, sheet metal material and labor in almost any fashion, greatly enhances control of each project discipline.

Figure 30–1
Industrial plant installation.
All Photography courtesy of Brud Bravera

"Danna Industries has invested over one (1) million dollars in developing one of the most innovative state of the art sheet metal plants. We are a full service mechanical contractor, who does not depend on a subcontractor to perform sheet metal work, which are often times over 50% of the mechanical system value. This investment increases productivity, saves time, reduces costs and enables us to meet any aggressive schedule to get the job done.

"Complete with modernized equipment, our sheet metal plant will meet any demand. With floor space of over 30,000 square feet we have designed our plant for the highest efficiency, including full CAD/CAM integration from layout to fabrication.

"Our technicians guide the installation for the proper start-up and commissioning of your system. This attention to detail results in superior system performance that will minimize callbacks. Supervised installation is the key to proper commissioning. As we start each system, we analyze the value of performance to the criteria of design. Proper commissioning leads to efficient energy consumption and extended life cycles of your system."[1]

Additional details about the company can be found in the project named Danna Industries.

project details

General and Architectural Information

Location: Various exhibit trade shows

Gross square footage: 100

New construction. The owner will build the exhibit in the company plant facility.

Codes and Governing Regulations

Design the project to comply with the most current edition of the International Building Code. Assume that the exhibit will be located in large assembly spaces such as convention centers and hotel ballrooms. Assume that these structures have a fire sprinkler suppression system. Select materials that will comply with the fire and safety requirements of these occupancy groups.

Facilities Required

TRADE SHOW EXHIBIT

Planning
- Assume that the exhibit will be contained in an area 10' x 10' x 10' high and bordered at the back with drapery. Plan for access from the front only.
- Photographs must be displayed on the exhibit. Provide for display of company brochures and business cards.
- Provide seating for customers.
- Provide a table or surface area for laptop computer slide show, gifts, and refreshments or candy. Gifts are company logo items, which can range from golf balls, leather portfolios, thermometers, clocks, baseball hats, polo shirts, jackets, and the like.

Design
- Incorporate a strong graphic identity with good visibility. The design should have a feeling of enclosure.
- Use materials related to DI's field of expertise. The exhibit will be constructed in the DI sheet metal plant.

---

1. Company brochure.

- Incorporate a lighting solution in the exhibit structure.
- The exhibit may not block view of adjacent exhibits on the right or left side.
- Your solution must incorporate the use of Figure 30–1 through Figure 30–9. The color images may be found on the CD and you may edit them in any way you choose.

Figure 30–2
Sheetmetal ductwork installation

Figure 30–3
Airport terminal installation

Figure 30–4
Airport terminal installation

Figure 30–5
Airport terminal installation

Furniture
- Seating for customers and a table
- Display fixture for company brochure and business cards

Equipment
- Electrical power duplex outlets, which are supplied to the exhibit space by the convention center
- Laptop computer

Figure 30–6
Industrial plant installation

Figure 30–7
Industrial plant installation

Figure 30–8
Research center piping installation

Figure 30–9
Piping installation

SUBMITTAL REQUIREMENTS

Phase I: Part A Due: _____   Part B Due: _____
Part C Due: _____   Part D Due: _____

A. Plan indicating any special design elements and attached interior fixtures keyed to details, sections, and elevations as required.

B. Elevations or details of exhibit showing design elements, fixtures, and color and material indications keyed to details and finish specification schedules.

C. Details showing sections through specially designed elements or fixtures must be prepared and identified as to wall locations.

D. Scaled model of exhibit. Write a brief description of how you have used three principles of design.

The plans should be presented in 1/4″ scale. The elevations are to be presented in 1/2″ scale. The model must be a minimum of 1″ = 1′ - 0″ scale.

Evaluation Document: Exhibit Design and Model Presentation Evaluation Form

Project Time Management Schedule: Schedule of Activities, Chapter 31.

Use the AutoCAD LT 2002 drawing file template named exhibit.dwg on the CD. It is a blank drawing template.

REFERENCES

Books

Buckles, G. Matthew. 1991. *Building architectural & interior design models fast.* Rancho Cucamonga, CA: Belpine Publishing Company.

Mitton, Maureen. 2003. *Interior design visual presentation: A guide to graphics, models, and presentation technique, second edition.* New York: Wiley.

Special Resources

Exhibitgroup/Giltspur: http://www.e-g.com
Think Exhibits: http://www.thinkexhibits.com

NOTES

SECTION

II

SVBMITTAL RESOVRCES

1'-2"

5'-9" @ GOWNS, COATS AND ROBES

4'-8" @ BLOUSE/PANT COORD'S, DRESSES AND JACKETS
4'-7" @ MEN'S PANTS (UNFOLDED)
4'-0" @ SHORT DRESSES

5'-9"

4"

3'-2"

1'-4"

3'-6"

SECTION ELEVATION

TYPICAL SINGLE
HANG BAR HEIGHTS

SECTION ELEVATION

TYPICAL HANGBAR
W/FACE OUT

2'-0"

3"

2'-0"

RECESSED STANDARD

VARIES

2'-0"

RECESSED STANDARD

SECTION ELEVATION

SECTION ELEVATION

CHAPTER 31

Submittal Documentation

EVALUATION FORMS

The submittal phase of each project includes a suggested evaluation document illustrated on pages 307 through 313. The forms are structured in relation to the scope of services of an interior designer and are expanded to include detailed design process tasks for each phase. Following this format supports the learning of project management skills. The broad format of a point system for each task is based on typical jury sheets used for student design competitions. The point allocation is to be determined by the design educator.

The top line of the form indicates what phase the form manages. Following is a line to indicate the project title and the date. The evaluation scale box is in the top left corner of the form. This is where the point allocation is to be indicated by the design educator. The adjacent long vertical boxes are provided for student names (the first box is to be left open).

A wide column is provided on the far left listing the tasks of the phase. The tasks for "other requirements" relate to those that do not occur in every project. Additional space is provided at the bottom should the instructor wish to insert additional items for evaluation. A line is provided for credit for work beyond minimum requirements, which is greatly encouraged.

The first narrow column can indicate the total points possible for the task on the left. The subsequent columns are provided to indicate the points earned by the student for the task. The final line is for remarks and the name of the individual performing the evaluation.

Oftentimes I had the students evaluate each other using these same forms. An outside jury is encouraged whenever possible. It is also good practice, time permitting, for the instructor to review the evaluation with the student before, during, and after the project presentation.

Remember that there is rarely one single resolution to a design problem and the solutions presented on the instructor's CD are for information only and are not intended to be the definitive answer. The electrifying part of teaching design is giving support to the delivery of individual student creativity and the exhilarating results.

Programming Evaluation Form

Project: _____   **Date:** _____

Evaluation Scale:

...............Points: ....
...............Points: ....
...............Points: ....
...............Points: ....
...............Points: ....

Programming

Research + Collect Data + Analysis	
Code Issues	
Matrix + Bubble Diagrams	
Prepare Core Plans + Elevations	
Visualization Sketches	
Concept Statement	
Preparation of Project Schedule	
Ability to Meet Schedule	

Other Programming Requirements

Role Playing	
Brainstorming	
Synectics	
Buzz Sessions	
Dialog Groups	
Team Participation and Responsibility	
Program Preparation	
Block Plans	
Interviews	

Bonus for work beyond minimum.

Total Points

Remarks

By: _____   **Page** ____ **of** ____

Note: Programming Phase is also referred to as Strategic Planning.

Conceptual Design Evaluation Form

Project: _____   **Date:** _____

Evaluation Scale:

............Points: ....
............Points: ....
............Points: ....
............Points: ....
............Points: ....

Conceptual Design

Preliminary Planning Concepts	
Preliminary Design Concepts	
Preliminary Furniture and/or Fixture Selection	
Preliminary Color + Materials	
Preliminary Ceiling Design	
Preliminary Lighting	
Preliminary Elevations	
Perspective Sketches	
Preliminary Budget + Cost Benefits	
Proper Plan Annotations	
Ability to Meet Schedule	

Other Conceptual Design Requirements

Charette Verbal Presentation	
Preliminary Power Plan	
Preliminary Floor Covering Plan	
Fire and Safety	
Accessibility	
Mock Preliminary Plan Check	
Schematic Fixture Design	

Bonus for work beyond minimum.

Total Points

Remarks _____

By: _____   **Page** _____ **of** _____

Note: Conceptual Design is also referred to as Schematic Design or Preliminary Design.

Design Development Evaluation Form

Project: _____ **Date:** _____

Evaluation Scale:
- ·········· Points: ········
- ·········· Points: ········
- ·········· Points: ········
- ·········· Points: ········
- ·········· Points: ········

Design Development

- Final Planning Concepts
- Final Design Concepts
- Final Furniture and/or Fixture Selection
- Final Color + Materials
- Final Ceiling Design
- Final Lighting
- Final Elevations
- Perspective Sketches
- Renderings
- Final Budget + Cost Benefits
- Ability to Meet Schedule

Other Design Development Requirements

- Does It Communicate the Concept and Program
- Ability to Be Built
- Solution Compatible with Building
- Solution Ecologically Responsible

Bonus for work beyond minimum.

Total Points

Remarks _____

By: _____ **Page** ___ **of** ___

Final Design Presentation Evaluation Form

Project: _____   **Date:** _____

Evaluation Scale:

   ....**Points:** ....

   ....**Points:** ....

   ....**Points:** ....

   ....**Points:** ....

   ....**Points:** ....

Presentation

- Planning Concepts Endorse Program and Research Data
- Design Concepts Endorse Program and Researched Data
- Egress, Fire and Safety, and ADA Code Compliance
- Furniture, Fixtures + Finishes Functional and Suitable for Use
- Creativeness of Planning Solution
- Rationality of Planning Solution
- Inventiveness of Design Solution
- Practicality of Design Solution
- Inventiveness of Furniture Fixtures and Materials
- Practicality of Furniture Fixtures and Materials
- Final Ceiling Plan Solution
- Final Lighting Plan Solution
- Elevations
- Renderings
- Budget + Cost Benefits
- Ability to Meet Schedule
- Oral Presentation
- Graphic Readability, Correctness (use of scale) and Clarity
- Proper Labeling and Keying
- Craftsmanship and Neatness
- Organized Presentation of Materials

Bonus for work beyond minimum.

Total Points

Remarks _____

By: _____   **Page** ____ **of** ____

Note: Evaluation elements are typical for ISP, ASID, and IIDA student design competitions and a few of those elements are incorporated above.

Construction Documentation Evaluation Form

Project: _____   **Date:** _____

Evaluation Scale:

................Points: ....
................Points: ....
................Points: ....
................Points: ....
................Points: ....

Construction Documentation

Title Sheet																
Wall Plan																
Furniture Plan																
Fixture Plan																
Ceiling Plan																
Lighting Plan and Switching Plan																
Power Plan																
Floor Covering Plan																
Elevations																
Sections																
Schedules																
Color and Material Specifications																
Furniture Specifications																
Graphic Readability, Correctness (use of scale), and Clarity																
Proper Labeling and Keying																
Craftsmanship and Neatness																
Organized Presentation of Materials																

Other Requirements

Bidding Documents																
Maintenance Schedules																
Post Occupancy Evaluation																

Bonus for work beyond minimum.

Total Points

Remarks _____

By: _____   **Page** _____ **of** _____

Design Agreements and Proposals Evaluation Form

Project: _____ **Date:** _____

Evaluation Scale:

.... Points: ....
.... Points: ....
.... Points: ....
.... Points: ....
.... Points: ....

Proposal Elements

Cover Page
Title Page and Table of Contents
Analysis of Problem and Statement of Concept
Pictorial Concepts
Project Methodology
Staffing and Schedule
Interior Costs
Design Firm Information and References

Agreement Elements

Dates, Identity of Client and Designer, and Project Description
Scope of Services
Fees, Terms, Reimbursable Costs, and Additional Services
Time Limit of Agreement and Party Responsibilities
Ownership and Use of Documents
Assignment, Arbitration, and Termination
Retainer
Signatures

Other Proposal and Agreement Requirements

Organized Presentation of Materials

Bonus for work beyond minimum.

Total Points

Remarks _____

By: _____   **Page** _____ **of** _____

Exhibit Design and Model Presentation Evaluation Form

Project: _____ **Date:** _____

Evaluation Scale:

.... Points: ....
.... Points: ....
.... Points: ....
.... Points: ....
.... Points: ....

Design Elements

Analysis of Problem

Inventiveness of Design Concept

Use of Materials

Graphic Design Concept

Communicates Identity of the Company

Presents Company Capabilities and Projects

Use of Principles of Design

Drawing Elements

Plan

Elevations or Details

Graphic Clarity

Three Dimensional Model Elements

Correct Use of Scale

Crafsmanship and Neatness

Representation of Materials

Communicates the Concept

Other Requirements

Ability to Meet Schedule

Organized Presentation of Materials

Ability to Be Built

Bonus for work beyond minimum.

Total Points

Remarks _____

By: _____ **Page** ___ **of** ___

CAD layer designations

I am grateful for CAD drawing technology, embrace all the timesaving features, and do not miss the tinge of graphite on my clothing, but the artist in me laments for the beauty of hand-drafted drawings. Ernie Sy and Benny Quan, my former cube' farm neighbors, were masters of the skill and I learned a great deal from them. Find a mentor like them and it will be much easier to look forward to the hand sketching and drafting that are still required in your conceptual design work. On the other hand, you are the designers of the future and the reason why all of the drawings provided on the CD are drawn with or converted to AutoCAD LT 2002 and are in .dwg format. In most cases, layouts are scaled to the required project scale and drawing page setups are defined to accommodate that scale. As a general rule you will find a layout tab for plan views, elevations, and ceiling plans. Plotting styles or dimension styles are not formatted in the CAD drawings on the CD.

Students are strongly encouraged to become familiar with the layer designations as the National CAD Standards and AIA define them. They are put into practice by many design firms for reasons of using a common drawing language and for exchanging drawing files between design consultants.

The layer designations in the drawings provided are simplified and loosely based on National CAD Standards and AIA Guidelines for the Interior (I) Layers. Architectural (A) layers are combined and simplified and finally designated in only four layers, which represent your floor outline for the projects. By no means are they static or all-inclusive, as I consider it not crucial to "micro manage" drawings on small projects such as those in this book. You, of course, may add the other layers as required for your projects, in particular layers for furniture systems.

Layer Name	Description
A-COLUMN-GRID	Column grid
A-SHELL	Building walls, windows, elevators, level changes, and ramps
A-SHELL-PATT	Wall and Column hatch patterns
A-STAIRS	Stairs and escalators
I-WALL	Interior walls
I-DOOR	Doors: swing and leaf
I-FLOOR-IDEN	Room numbers, names, etc
I-FLOOR-CASE	Casework
I-FLOOR-PATT	Paving, tile, carpet, etc
I-FLOOR-FIXT	Store fixtures, display cases, miscellaneous fixtures, etc
I-EQUIP	Equipment
I-FURN	Furniture
I-FURN-IDEN	Furniture numbers
I-CLNG	Ceiling
I-LITE	Light Fixtures
I-ELEV	Elevations
I-SECT	Sections
I-DETL	Details
E-POWER	Electrical receptacles, junction boxes, telephone, cable, etc
E-SWITCH	Light switch
I-BORDER	Drawing Border
I-PERIM-FIXT	Perimeter wall store fixtures, shelving, cases, etc

Table 31-1 Interior Layer Designations

Three projects in this book (pma.dwg, chippewa.dwg, and joie.dwg) follow an architectural layer designation format that has more levels of layer separations for drawing elements. Upon opening one of the three drawings or reviewing the designations below, you see that up to four layers put together a door. When you are finished walking all the way through this mini labyrinth, you are required to add the interior layers to these projects. This is not a specific assignment but a general task for each project without interior layer designations. An architectural office provided these drawings and the layer designations follow.

Layer Name	Description
AN-ANNO	Annotations
AN-ANNO DIM	Annotations, Dimensions
AN-ANNO ELEV	Annotations, Elevations
AN-ANNO LDR	Annotations, Leaders
AN-ANNO RM#	Annotations, Room Numbers
AN-DEMO	Demolition
AN-DOOR	Doors
AN-DOOR ANNO	Door Annotations
AN-DOOR JAMB	Door Jambs
AN-DOOR SWING	Door Swings
AN-FLOR ABOVE	(Bulkheads — ceiling reference)
AN-FLOR APPL	Floor Appliances
AN-FLOR BCABS	Floor Base Cabinets
AN-FLOR HRAIL	Floor Handrails
AN-FLOR PFIX	Floor Plumbing Fixtures
AN-FLOR ROD	Floor Rod
AN-FLOR SHELF	Floor Shelf
AN-FLOR STAIR	Stairs
AN-FLOR STAIR ANNO	Stair Annotations
AN-GLAZ	(Windows)
AN-GLAZ JAMB	Glazing Jamb
AN-GLAZ SILL	Glazing Sill
AN-WALL	Wall
AN-WALL PATT	Wall Pattern
AX-*****	
ELEV-OO HATCH	Elevation # Hatch
ELEV-OO	Elevation #
F-PROT	Fire Protection
F-PROT ANNO	Fire Protection Annotate
L-SITE CURB	Landscape Site Curb
L-SITE DECK	Landscape Site Deck
L-SITE PATT	Landscape Site Pattern
L-SITE WALK	Landscape Site Walks

A=Architectural+

    D=Demo (AD)

    N=New (AN)

    X=Existing (AX)

Table 31-2 Architectural Layer Designations

It was decided to allow you the freedom to suggest locations for HVAC registers and sprinkler heads as they work with your designs for the sole purpose of getting you used to the idea that they are always going to be part of the building system. (Do not forget, "these things" can be painted to match the ceiling, but no wallcovering please!) In most cases, designers are consulted regarding their final placement relative to special design elements. Required NFPA fire protection sprinkler devices are well defined in the books noted at the end of this chapter. The following is a rule of thumb for HVAC:

- Office Spaces: 1,600 square feet equals one zone. Locate three to five registers in a zone. The ratio is two supply registers to one return register.
- Retail Spaces: 1,800 square feet equals one zone. Locate three to five registers in a zone. The ratio is two supply registers to one return register.

REFERENCES

Resources: Metric Conversion Software, Versaverter 2.0, developed by Scott Wayne Baker (PawPrint.net) http://www.pawprint.net.

Fire Protection: Ballast, David Kent. 1994. *Interior construction and detailing for designers and architects*. Belmont, Calif.: Professional. Harmon, Sharon Koomen, and Katherine E. Kennon. 2001. *The codes guidebook for interiors*. New York: Wiley.

CAD Management: Grabowski, Ralph. 2002. *CAD manager's guidebook*. Albany, N.Y.: OnWord Press.

Drafting: Kilmer, W. Otie and Rosemary. 2003. *Construction drawings and details for interiors: Basic skills*. New York: Wiley.

project management forms

Project Record Sheet
Project:

Room Name

Flooring	*(e.g. Stark. Field, Madison, Red, 27" w. 100% worsted wool)*
Transition Materials	
Base	
Trim	
Architectural Details	
Columns	
Stair Elements	

Walls	*(e.g. Pallas Walls. Bird's Eye. Cremo. #28-002-039.*
	Accent wall: Silk Dynasty, Palazzo Collection, #F-7, Florence)

Switch Plate + Register	
Casework	

Fixtures	

Hardware	
Plumbing Fixtures	
Plumbing Hardware	
Ceiling	
Ceiling Light Trim	
Ceiling Register Trim	
Lighting	
Window Covering	

Furniture	*(e.g. Jeffco. #9417. Ottoman. 33 x 33 x 18 h. #62 Palm Crackle Finish. (2). Fabric.*
	Scalamandre. #36049-007.)

Appliances	
Equipment	
Other	

Design Firm Logo and Address

FURNITURE SPECIFICATION

PHOTO	DATE ISSUED
	REVISED △
	REVISED △
	PROJECT
	ROOM
	ITEM NO **F . #**
	PREPARED BY

	DESCRIPTION	QUANTITY	UNIT PRICE	EXTENDED PRICE
MANUF				0.00
				0.00
				0.00
				0.00
MODEL NO				0.00
				0.00
				0.00
DIMENSIONS W D HT DIAM				0.00
SEAT HT ARM HT				0.00
FINISH				0.00
				0.00
UPHOLSTERY	YDS____X____=	YDS		0.00
				0.00
COM MANUF				0.00
				0.00

NOTES		TOTAL LIST	$0.00
VENDOR		DISCOUNT	$0.00
DELIVERY		FREIGHT	$0.00
		SUBTOTAL	$0.00
INSTALLATION		TAX	$0.00
		TOTAL NET	$0.00

Prices are for budgeting purposes only and are not guaranteed. Purchaser to verify prevailing or future prices.

F . #

[1]

Color + Material Documentation   **PAINT**[2]

Project No.:

Project:

Page: *of*

Date Issued:

Revisions:

△
△   Date:[3]
△   Date:
       Date:

Manufacturer No. + Color [4]	*Sample*[5]	*No.*[6]
PPG #2550 Navajo White		**PT1**
Design Tex, "Reno", #6340-103 100% Olefin. Acrylic backing. Scotch guard finish. 54" Wide. Repeat: V-1 5/8" H-3 9/16".		**WC1**

---

[1] Design Firm logo, address, contact numbers, etc.
[2] Indicates type of material such as paint, laminated plastics, carpet, marble, wallcovering, etc. For large projects a separate sheet would be used for each type of material.
[3] Revision dates.
[4] Indicate complete information about material that would be necessary for budgeting, bidding, or purchasing process.
[5] Sample of material is attached below. For heavier materials cover stock would be used for the page.
[6] Number or letter code that appears on the finish plans or elevations in the construction document drawings.

Design Firm Logo and Address

PURCHASE ORDER: CT4499

Purchase order number must appear on all forms relating to this order.

DATE

TO: DONGHIA TEXTILES
485 BROADWAY
NEW YORK, NEW YORK 10013
212-925-2777
ATTN: ORDER DEPARTMENT
RESERVED 1.22.48765

P.O. NUMBER CT4499
REVISED P.O. DATE
REQUISITIONED BY
SHIP BY
SHIP VIA BEST WAY
F.O.B.
TERMS
OTHER

SHIP TO:
BAKER FURNITURE COMPANY
2219 SHORE DRIVE
HIGHPOINT, NC 27263

QTY	UNIT	DESCRIPTION	PRICE	AMOUNT
8.5	YDS	DONGHIA #7460-04. KALEIDOSCOPE, COLOR MINERAL, 56"WIDE, 12'V, 6"H 40%VISCOSE, 36%COTTON, 24%LINEN. (RESERVED) **TAG: PO#27559**	68.00	578.00
4.5	YDS	DONGHIA #7460-03. KALEIDOSCOPE, FOOL'S GOLD, 56"WIDE, 12'V, 6"H 40%VISCOSE, 36%COTTON, 24%LINEN. (RESERVED) **TAG: PO#27559**	68.00	306.00

Terms:

(Financial terms of purchase. Detailed terms are typically indicated on back.)

SUBTOTAL	884.00
FREIGHT	20.00
TAX RATE	7.000%
TAX	63.28
TOTAL DUE	$967.28

Authorized by   Date

Logo and address

27 November 2006

INVOICE   No.

Client Name
(address)
(address)
(project number)

Time Summary

Task	Hours	Description	Rate	Amount
Planning + Programming				
Design				
Construction Documents				
Clerical				
Meetings				
Other				

Purchase Orders

(list p.o. numbers)

Purchasing Fee

Reimbursables

Telephone		
Fax		
Copies		
Blueprints		
Other		

	Subtotal
	Tax Rate
	Tax
	Total Due

Past-due invoices are subject to finance charges of   % per month (% annually) on the balance due.

Due upon receipt.
Thank You.

REFERENCES
Books

Crawford, Tad. 2001. *Business and legal forms for interior designers.* New York: Watson-Guptill.

sample matrix

Additional information on this planning tool can be found in Designing Interiors, Interior Design, Second Edition and many other interior design textbooks.

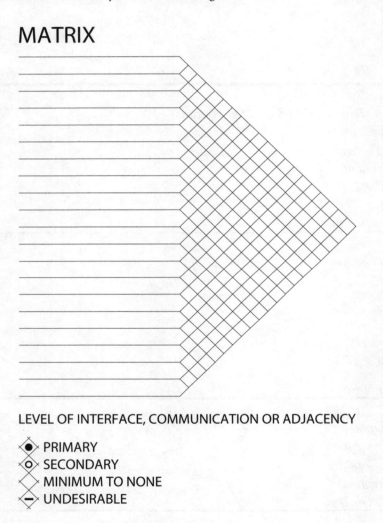

MATRIX

LEVEL OF INTERFACE, COMMUNICATION OR ADJACENCY

- PRIMARY
- SECONDARY
- MINIMUM TO NONE
- UNDESIRABLE

time management + project schedule form

A project schedule is an invaluable tool for helping you to manage your time. It is used extensively in design offices to manage the time involved to complete projects and meet project deadlines. You are required to prepare this schedule for each project excluding the project charettes. You can also manage all your course work by making a new table in which you would insert the work that has to be completed in the left column.

Schedule of Activities

Project: _____   **Date:** _____

Week or Month																								
Date																								

*

Programming
Research + Collect Data
Matrix + Bubble Diagrams
Preliminary Planning Concepts
Prepare Core Plans + Elevations
Block Plans
Space Plans

Schematic Design
Preliminary Furniture Selection
Preliminary Color + Materials
Ceiling + Lighting Design
Preliminary Budget

Design Development
Finalize Furniture
Finalize Color + Material
Finalize Ceiling + Lighting Design
Finalize Plans + Elevations
Prepare Final Budget

Presentation
Detail, Cross-Reference + Print Plans
Collect Material Samples + Furniture
Mount Plans + Samples
Label Plans + Boards
Prepare Agenda
Review Evaluation Forms

Phase: _____   **By:** _____   **Page 1 of 2**

Schedule of Activities

Project: _____ Date: _____

	Week or Month														
	Date														
Construction Documentation *															
Partition Plan															
Furniture Plan															
Fixture Plan															
Elevations															
Floor covering Plan + Details															
Details + Sections															
Reflected Ceiling Plan															
Lighting Criteria Plan															
Electrical Power Criteria Plans															
Floor Covering Plan + Details															
Details + Sections															
Color and Material Schedules															
Furniture Specifications															
Fixture Specifications															
Title Sheet															
Bidding Documents															
Maintenance Schedules															
Post Occupancy Evaluations															

Phase: _____   By: _____   **Page 2 of 2**

typical retail drawings

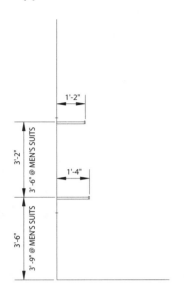

SECTION ELEVATION
TYPICAL DOUBLE
HANG BAR HEIGHTS

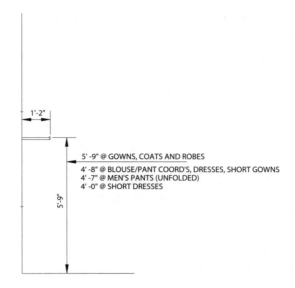

5'-9" @ GOWNS, COATS AND ROBES

4'-8" @ BLOUSE/PANT COORD'S, DRESSES, SHORT GOWNS
4'-7" @ MEN'S PANTS (UNFOLDED)
4'-0" @ SHORT DRESSES

SECTION ELEVATION
TYPICAL SINGLE
HANG BAR HEIGHTS

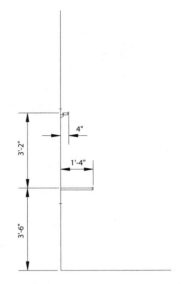

SECTION ELEVATION
TYPICAL HANGBAR
W/FACE OUT

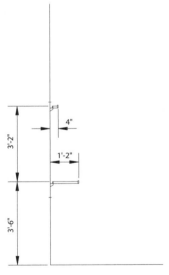

SECTION ELEVATION
TYPICAL DOUBLE FACE OUT

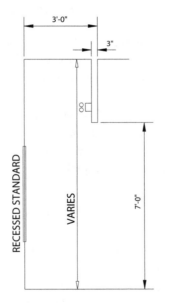

SECTION ELEVATION
TYPICAL VALANCE OR CURTAIN WALL
FOR HANGING MERCHANDISE

SECTION ELEVATION
TYPICAL VALANCE OR CURTAIN WALL
FOR SHELVING

Figure 31–15
Fixture dimension criteria for perimeter walls

government regulations

ADA

How many of you have participated in one of the most basic (or should I say historic) situational research assignments of being sent off for a day blindfolded or in a wheelchair or on crutches? Had you taken the challenge, many of the ADA requirements would be simple commonsense decisions for you. There are so many great resources already in print so it would be superfluous to reprint the graphic illustrations for accessible design. Hopefully, these resources are most likely found in your personal or school design library. If not, obtain the ADA Standards for Accessible Design generously made available in PDF format from the United States Department of Justice.

The following is a list of resources to assist you in planning user-friendly design for your projects and for compliance with the Americans with Disabilities Act.

INTERIOR DESIGN TEXTBOOKS

The following books are dedicated solely to the subject and presented in illustration format:

Leibrock, Cynthia, and James Evan Terry. 1999. *Beautiful universal design: A visual guide.* New York: Wiley.

Leibrock, Cynthia, and Susan Behar. 1997. *Beautiful barrier free: A visual guide to accessibility.* Second edition. New York: Wiley. (QA 5)

The above books have specific chapters or appendixes that discuss or graphically illustrate universal design and barrier free issues:

Ballast, David Kent. 1994. *Interior construction and detailing for designers and architects.* Belmont, Calif.: Professional.

DiChiara, Joseph, Julius Panero, and Martin Zelnik. 2001. *Time-saver standards for interior design and space planning.* 2d ed. New York: McGraw-Hill.

Kilmer, Rosemary, and W. Otie. 1992. *Designing interiors.* Forth Worth, Tex.: Harcourt Brace Jovanovich College Publishers.

Pile, John F. 1995. Interior design. 2d ed. New York: Abrams.

Reznikof, S. C. 1986. *Interior graphic and design standards.* New York: Watson-Guptill.

Reznikof, S. C. 1999. *Specifications for commercial interiors: Professional liabilities, regulations, and performance criteria.* New York: Watson-Guptill.

UNITED STATES DEPARTMENT OF JUSTICE

"ADA STANDARDS FOR ACCESSIBLE DESIGN are in Appendix A of the Title III Regulations.

ADA Standards for Accessible Design (Acrobat PDF format) (4.5 MB file).

http://www.usdoj.gov/crt/ada/stdspdf.htm

The PDF version of the ADA Standards contains the full formatted text and graphics, as published in the Code of Federal Regulations, complete with links to figures, graphics and cross-referenced sections. The file may be downloaded and saved to a computer and once this is done, it may be viewed without having to use the Internet.

ADA Standards for Accessible Design (HTML format).

http://www.usdoj.gov/crt/ada/reg3a.html#Anchor-Appendix-52467. This version of the Standards is in HTML format. It has links to the figures from the Standards but the images do not appear on the same page as the related text. The figures appear in a separate window. Text descriptions of the figures are also included."[1]

1. U.S. Department of Justice. ADA Publications.

Building Codes

In this book the intent of the code references is that you learn to act and decide responsibly in addressing life safety issues related to interior planning and design such as egress, accessibility, and finishes. Keep that in mind eternally. The motives of protecting the health, safety, and welfare of the public are the major goals of construction standards. At all times I choose to err on the side of caution, especially when it comes to selecting materials for projects, even if the code allows for a lower rating because of occupancy or sprinkler systems. The materials with the highest rating are typically more expensive, but when it comes to saving a life, I judge it well worth it.

The project exercises also introduce the students to the many model codes used in different jurisdictions. It is up to the judgment of the instructor to decide which codes to apply to the project exercise. For instance, you may opt to adhere to the IBC, or a building code that the students are familiar with, or your local adopted codes for the projects. Governmental building and design codes vary and model building codes are in a state of transformation, which fundamentally means that the information on the drawings or in the project details cannot be held by the reader as absolute or up-to-the-minute information. Although every attempt was made to be accurate during the writing of this book, in some cases, the codes used when the project was completed are no longer in effect. The jurisdictions that use the Uniform Building Code (UBC) today could also be in the process of adopting the International Building Code (IBC). In the long run, it will be easier for the design industry when the International Code Councils' goal of developing a single set of comprehensive and coordinated national model construction codes reaches full fruition and the international codes are adopted throughout the country.

REFERENCES

For a complete list of the International Code Adoptions by State and Jurisdictions, go to the International Code Council, Inc. at http://www.iccsafe.org/government/adoptions.htm. The tables include reference to the International Building Code (IBC), ICC Electrical Code (ICCEC), International Energy Conservation Code (IECC), International Fire Code (IFC), International Property Maintenance Code (IPMC), International Fuel Gas Code (IFGC), International Mechanical Code (IMC), International Plumbing Code (IPC), International Code (IPMC), International Residential Code (IRC), International Private Sewage Disposal Code (IPSDC), International Zoning Code (IZC), International Performance Code for Buildings and Facilities, and International Urban-Wildland Interface Code (IUWIC).

The Council of American Building Officials (CABO) document Introduction to Model Codes is available for download in Adobe Acrobat .pdf format at: http://www.iccsafe.org/news/guides.htm. It is a thirty-one page publication with a basic explanation of the nationally recognized model code organizations and the codes and standards development procedure. It also lists the building code organizations addresses and Web sites.

A final caution note must be clear. All drawings and project exercises are presented solely to provide tutorial information about interior design processes. The information is to be used for those purposes only, and it is not intended for construction or otherwise. These drawings can be used with the understanding that the authors are supplying information but are not attempting to render architectural, engineering, or other professional services. If such services are required, the assistance of an appropriate professional should be sought. The information contained in the drawings is provided on the basis of general knowledge of the interior design industry. Governmental building and design codes vary, and none of the information on the drawings can be held by the reader as legal information. The authors will not assume any responsibility for designs based on information contained on these drawings. No warranty or fitness is implied. The authors shall have neither liability nor responsibility to any person or entity with respect to any loss or damage in connection with or arising from the information contained on these drawings.

INDEX